Golden Rocks: The Geology and Mining History of Golden, Colorado

By
Donna S. Anderson and Paul B. Haseman

©2021 Donna S. Anderson and Paul B. Haseman

ISBN: 979-8-9850417-2-9 (Softcover)
Golden, Colorado, USA

Please send all comments and corrections to
goldenrocksgeology@gmail.com

Original Cover Art ©2021 Jesse Crock.
Photographs and figures by persons other than the authors are credited in the figure captions.
Sketches redrawn by Vivian Haseman.

Table of Contents

Table of Contents i
Preface iv
Abbreviations v
Chapter 1: Golden Rocks 1
    Golden's Unique Setting 1
        Landforms and Geology 3
        Golden's Geologic Rosetta Stones 3
        Organization 5
Chapter 2. Primordial Beginnings 6
    Front Range Metamorphic Bedrock 7
    Golden's "Red Rocks" 8
        Fountain Formation 8
        Lyons Formation 9
        Lykins Formation 9
Chapter 3: Jurassic Park and the Western Interior Seaway 12
    Golden's Jurassic Park 13
    Shorelines of the Western Interior Seaway 14
        Crawling with Ancient Life 15
    Underwater for a Long Time 15
        Benton and Niobrara Ocean Worlds 16
        Pierre Ocean World 16
Chapter 4: Good-bye Seaway, Hello Mountains 18
    Beaches and Swamps 19
        Fox Hills Beach 19
        The Swamp-Infested Laramie Formation 19
    Volcanoes and Rushing Rivers: The Rockies at Golden's Doorstep 22
        A Rising Mountain Range: The Arapahoe Formation 22
        Denver Formation: Mass Extinction and Lava Flows 23
    Fans of Green Mountain 28
Chapter 5: Mountains UP! 29
    Squashing Golden 30
        Severe Squashing 30
        Gentler Squashing 32
        Below the Ground 33
        After 64 Ma in Golden 34
        Ultimate Cause of Squashing 35
    Modern Earthquakes in Golden? 35
        National Earthquake Information Center 36

Chapter 6: Rockies Rebirth 37
Clear Creek Shapes the Golden Landscape 37
Ancestral Clear Creek 38
Ice-Ages and Clear Creek 38
Channeling Clear Creek 43
Historic Flows and Floods 44

Chapter 7: Water is Golden 46
Resourceful Clear Creek 46
Whiskey is Made for Drinking 47
Tales of Metals and Nutrients 50
Water Below the Ground 53
Golden Water Wells 55
Springing Forth 56

Chapter 8: Going for the Gold 58
Gold in Golden? 58
Golden's One Gneiss Mine 60
Smelters in Golden 61
CSM Experimental Plant (CSMRI) 62

Chapter 9: Black Diamonds 63
Coal Below the Ground 63
Golden Coal Mines 64
1889 White Ash Mine Disaster 67

Chapter 10: From a Lump of Clay 69
Clay's Anatomy 69
Clay Stoping and Open-Pit Mining 69
Golden Clay Mines 70
Dakota Fire-Clay 71
Laramie Brick-Clay 73

Chapter 11: No Stone Unturned 77
The Bastard Limestone 78
Cambria Lime Kiln 78
Sand and Gravel Quarries 79
Gneiss Quarries 80
Basalt Quarries 81
North Table Mountain: From Zeolites to Riprap 81
South Table Mountain: The WPA Quarries 83

Chapter 12: Mining Legacies 86
Mining-Related Historic Rail Lines 86
Mine Subsidence 89
Mining History and Open Space 91
Collisions 92
Future Musings 93

Acknowledgments	95
Glossary	96
Geology Primer	104
Three Types of Rocks	104
Sedimentary and Structural Geology: There are Rules!	105
The Rock Cycle and Plate Tectonics	107
Geologic (Deep) Time	108
A Philosophy of the Rock Record	109
The Present and the Past: Uniformitarianism and Catastrophism	109
Reading Golden's Geologic Map	110
Reading Golden's Geologic Column	111
References Cited	113
About the Authors	121
Donna Anderson	121
Paul Haseman	121

Preface

An email from Paul Haseman in Fall of 2018, asking whether there was a guide to the geology of Golden, started the journey down the road to writing this book. As it turned out, the last geology guide, summarizing geologic knowledge about Golden, was written in 1938.

You might think that the rocks have not changed in the last 80+ years, but that is not strictly true. Geologic processes keep altering the rocks. Mining altered the rocks and in some cases uncovered them. Advancing urbanization is covering the rocks up and, in some cases, removing them. Geologic, including paleontologic, resources are disappearing. And, since 1938, geologists have found out a lot more about Golden's rocks and their origins. Even geologists are surprised to find out that quite a few of Golden's rocks were turning points in understanding fundamental concepts worldwide!

Our research made us realize that Golden has a rich mining history. Few books recount that history from a geologic viewpoint, particularly within Golden. More and more Golden residents, new to the area, are surprised to find out that Golden was a mining town in its own right from 1870 through the 1950s. Not only does geology shape the Golden landscape, but mining has shaped Golden, not least of which includes open space in and around Golden. The story of water in Clear Creek is another matter that is hard to find in one place. Clear Creek is a major attraction for Golden and Denver residents, and it is the source of Golden's good drinking water. Mining and water are inseparably intertwined. It is long past time to write about these issues.

We thought long and hard about how to write this book. A guiding concept was to make the geology accessible to the non-geologist, targeting a reader interested in science and history but who is not necessarily a professional scientist. Part of our inspiration came from Alan Alda's book on science communication: "If I understood you, would I have this look on my face?"

Every profession has its jargon, and geology is no exception. That jargon, however, becomes vocabulary that creeps into newspapers, social media, and other communication pathways. To that end, we have provided a Glossary and a short Geology Primer that attempt to explain some of the words and concepts unique to the field of geology. In our writing, we tried to use "plain," or conversational, English rather than the stilted, dense style of technical communication.

We chose to make this an open-access electronic publication. We want you to freely read and use the information in the book. If you want to print a copy, you are free to do so without restriction. If you want to use figures or quotations in something that you are writing or presenting, just give us the proper credit. But note that many photographs in the book are used with permission from other sources: you need to get their permission to reproduce them.

Your comments and corrections are highly valuable to us. In reseaching the mining history of Golden, we found conflicting and vague information. We have done our best to present the most solid data that we could find, but inevitably, errors have crept in. Let us know what you know via email: goldenrocksgeology@gmail.com. Electronic publications are much easier to revise than paper-copy books!

We hope that this book encourages you to understand the landscape beneath your feet and at the horizon by learning the stories on how that landscape came about. We also hope that you gain insight into the changing mindset of how humans have related to the land in the modern story of mining in Golden since 1859. Whether we realize it or not, those activities and attitudes still shape issues in our community today.

Donna Anderson and Paul Haseman
Golden, Colorado
October 2021

Abbreviations

aka "also known as"
C-470 Colorado Highway 470
CCRR Colorado Central Railroad
CDOT Colorado Department of Transportation
CDPHE Colorado Department of Public Health and Environment
cfs cubic feet per second
CSM Colorado School of Mines, now referred to as "Mines"
CSMRI Colorado School of Mines Research Institute
CT Colorado Transcript (became the Golden Transcript in 1969)
DPL WHC Denver Public Library, Western History Collection
e.g. "for example »
et al. "and others"
Fig., Figs. Figure, Figures
Fm., Fms. Formation, Formations
GIS Geographic Information System, basically a spatial data mapping system
GP Group
GPS Global Positioning Satellite
Hwy Highway, State or US
I-70 U.S. Interstate 70
Ibid. "the same reference as in the footnote above"
Jeffco Jefferson County, Colorado
K-Pg Cretaceous-Paleogene boundary
K-T Cretaceous-Tertiary boundary (now superceded by K-Pg)
Ma Million Years, from the Latin: mega-annum
MMM Martin Marietta Minerals, Inc.
Mt Mountain
NNL National Natural Landmark
US, U.S. United States
USGS United States Geological Survey
YPM Yale Peabody Museum

Chapter 1: Golden Rocks

Have you ever wondered how the landscape around the City of Golden came to be? Why are North and South Table Mountains so flat? What's up with all those rock fins on the back-nine at Fossil Trace Golf Course? What are those white stripes along the sharp ridge west of State Hwy-93 to Boulder? Were there ever any dinosaurs in Golden? Where did those round rocks in the Armory Building come from? Did anyone ever find any gold in Golden? Did you know that Golden was a mining town from 1870 through the 1950s? And, that most of the public open space in and around Golden has its roots in mining?

This book answers these questions and more by unraveling the tapestry of the Golden landscape. You will discover the origins of what you are walking or cycling by, driving across, golfing over, or simply just seeing every day when you walk out of your home. You will learn the story behind Golden's good drinking water. You will learn how Golden's early settlers used their local rock materials to build Golden and Denver, a legacy that still provokes controversy.

Golden's Unique Setting

Golden lies in a confined valley between mountainous geologic features that mark the transition from the Great Plains to the Rocky Mountains (Fig. 1). No other setting is like it along the Front Range of northeastern Colorado. Native Americans lived, hunted, and parlayed here for thousands of years,

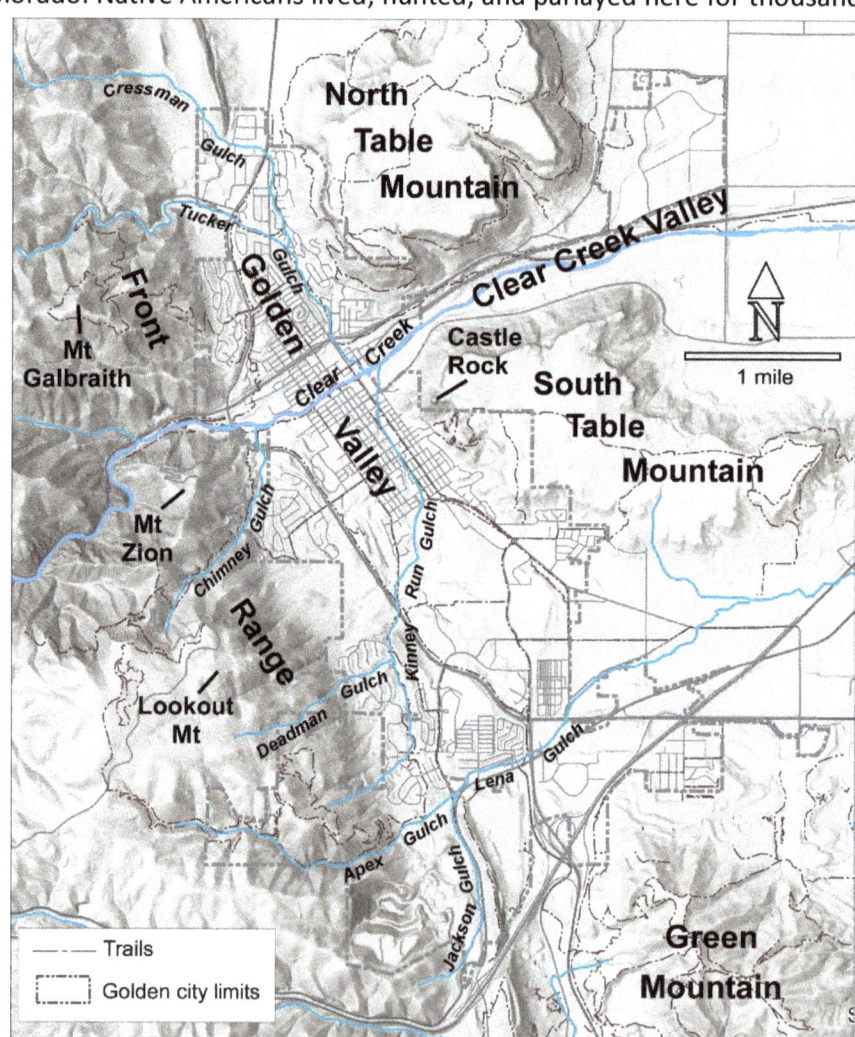

Fig. 1. Golden's landscape features. Basemap used throughout book. GIS data sources include City of Golden, CDOT, Jefferson County Colorado, State of Colorado data. Hillshade basemap from ESRI Online world imagery.

because the Golden Valley offered an outstanding sheltered place to live, with abundant game and water.[1] In the middle 1800s, Euro-Americans came to the valley with an eye toward developing Golden's geologic resources. Today, we appreciate Golden's setting for its open-space areas, trails, and views. Golden's geologic features set it apart from Denver as an island in a sea of urban sprawl.

Iconic North and South Table Mountains first meet the eye in Golden's landscape (Fig. 1). Rising above the Golden Valley, they form the sentinels separating Golden from the metropolitan area to the east. Castle Rock,[2] forming part of Golden's city logo, stands 650 feet above the Golden Valley floor and dominates eastern city views. At the south end of the city, Green Mountain rises nearly 1000 feet to its summit above the surrounding land, forming another iconic feature.

On the west side of Golden, the Rocky Mountain Front Range climbs abruptly up to a ridgeline 1500 feet above the valley floor (Fig. 2). Prominent features along the ridgeline are Mount Zion[3] with

Fig. 2. The Front Range backdrop of western Golden consists of Lookout Mountain, Mount Zion, and Mount Galbraith. Dashed white line is the approximate edge of the Rocky Mountains. Note the deep V-shaped Clear Creek Canyon. Colorado School of Mines campus in foreground. View to west: north is to right.

the large M (for Colorado School of Mines) that is lit up at night,[4] Lookout Mountain, with its tall antennas, and Mount Galbraith.[5] The ridgeline borders the high (7500 to 9000+ foot elevation) mountainous upland that rises west to the Continental Divide.

Clear Creek, now a major recreation destination, flows west to east across the Golden Valley (Fig. 1). Over geologic time, Clear Creek incised down through the Front Range Mountains (Fig. 2) on the west and between North and South Table Mountains on the east, forming a deep canyon on either side of the city. Within the city, Clear Creek broadens out forming its own valley, where Euro-American settlers founded Golden in 1859. Golden was a perfect site at the foot of the Rockies, with a permanent water source and near to trails leading to the western gold fields.

---

[1] Ute Mountain Utes, Southern Utes, Arapaho, Cheyenne, and earlier indigenous peoples flourished in the area before Euro-American settlers displaced them.
[2] Not to be confused with the town of Castle Rock, 32 miles southeast of Golden, in Douglas County, Colorado.
[3] Edward Berthoud called the high peak west of town "Mount Zion" because of the "difficulty in climbing it." Golden Globe 18 July 1874.
[4] The large "M" on Mt. Zion was put in place on 15 May 1908 by a crew of CSM faculty and students, who hiked up Mt. Zion with whitewash and tools strapped to several burros. Originally a 3-D geometry project of students and faculty, the M was lit up at night beginning 19 March 1932. Replenishing rocks on the M is a tradition carried out enthusiastically by CSM first-year students each Fall.
[5] Mt. Galbraith was named in 1977 in honor of Denman Galbraith (1918-1974), a Golden historian and former tank commander under General Patton at the 1944 Battle of the Bulge. Galbraith served on the Board of Foothills Art Center. (CT 16 February 1977 and USGS Geographic Names Information System database, #196056).

Landforms and Geology

Fig. 3. Yesterday (left) and today (right): Castle Rock and South Table Mountain from the 1869 Hayden party campsite (left). Sketch by H.W. Elliot in McKinney (2003). Views to southeast, taken from New Loveland Park.

On 14 July 1869, Dr. Ferdinand V. Hayden (1829-1887), a famous geologist and explorer of the Western U.S.,[6] came to Golden on horseback. He camped with his small field party just north of 8th Street between Illinois and Cheyenne Streets, with an excellent view of Castle Rock (Fig. 3). While exploring the area, he noted the prominent "metamorphic[7] foothills" and discussed at length the lava flows of the two Table Mountains.[8] His observations of dramatic landforms echo those of most people who come to Golden for the first time. But, as a geologist, he further recognized that the landscape was the direct result of the geology of Golden. Hayden noted how the rocks were oldest on the west, starting at the Front Range, and youngest on the east at the two Table Mountains. He also remarked on the flat rocks of the Table Mountains versus the tilted rocks on the west side of town.

What is so special about Golden's geology, aside from dramatic landforms? Hidden in its rocks, Golden has one of the best-preserved stories of how the Front Range of the Rocky Mountains evolved over the last 1700+ million years. That is a long time. The rocks tell stories of three mountain-building episodes; two mass extinctions of life on Earth; dinosaurs and woolly mammoths walking across the land at different times; lava flows running across the land surface; long periods of being under oceans; tropical to subtropical climates; and massive erosion and glacial flooding. Not only that, 19th century settlers exploited Golden's rocks to start industries, some of which have fallen away, while others still flourish. Once you adjust your eyes to the geologic underpinnings of the landscape, those stories come alive. To a geologist this is why "Golden rocks!"

Golden's Geologic Rosetta Stones

Golden's geologic map and geologic column (Fig. 4) are the Rosetta Stones for explaining the geology of Golden. Both show the names and ages of the rock formations that are present across Golden, but they show it in different ways. The geologic map shows the distribution of rock formations on the ground surface, along with geologic fault lines. At any location on the geologic map, you can tell the rock formation name and age at that spot by its color. For your phone or tablet you can download an app for that in the form of Rockd,[9] which has a GPS locator showing where you are standing and a description of the rock formation beneath your feet.

---

[6] F.V. Hayden also helped establish Yellowstone as the first National Park in the U.S. in 1871.
[7] Throughout the book, words highlighted in blue are defined in, but are not hyperlinked to, the Glossary.
[8] Hayden, 1869, p. 34-35.
[9] **Rockd**, created by the Macrostrat Lab at the University of Wisconsin–Madison: https://macrostrat.org/ is a free app.

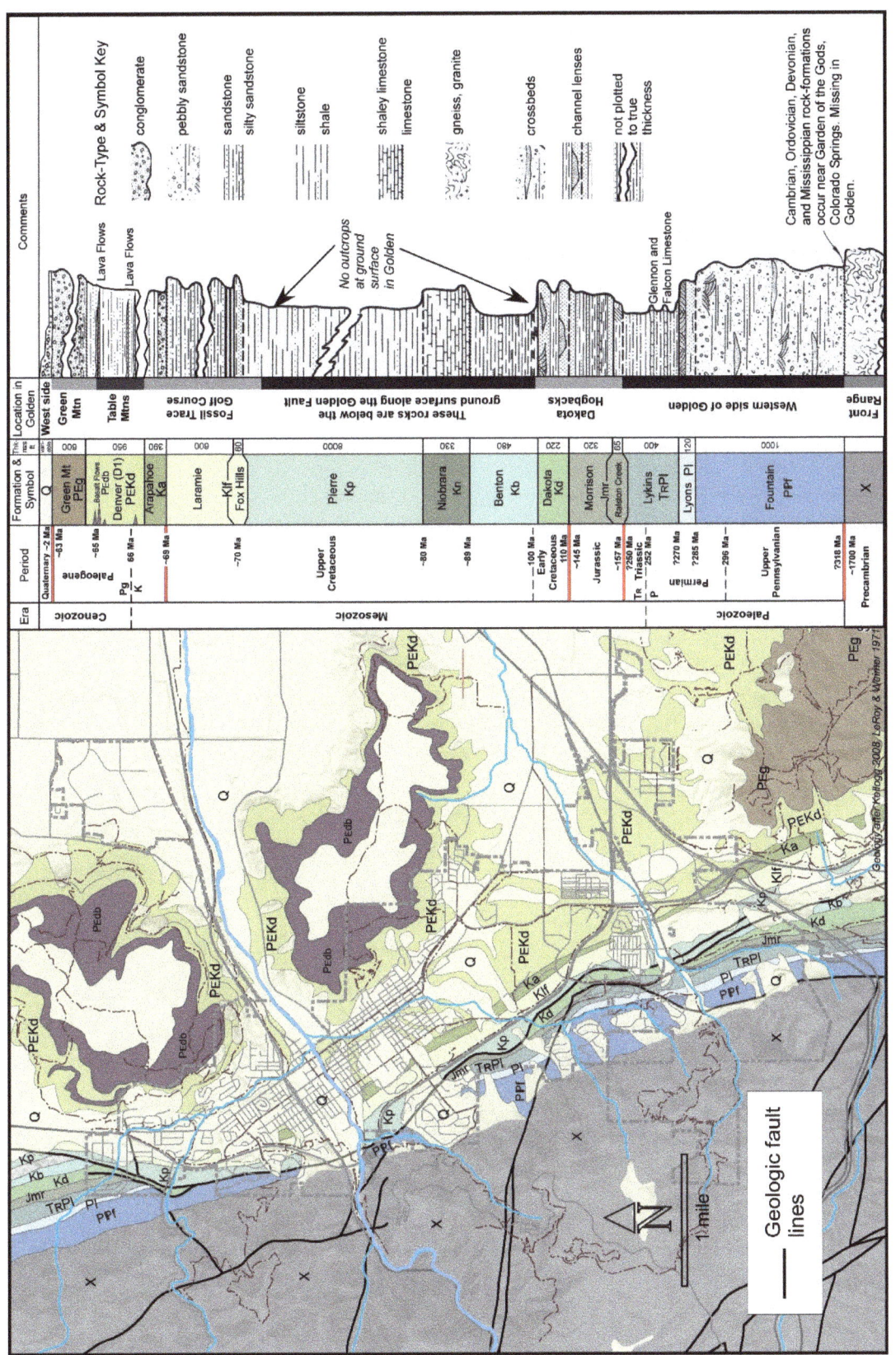

Fig. 4. Geologic map (left) and geologic column (right) for the Golden area. Basemap is the same as that in Fig. 1. Compiled from Kellogg et al. (2008) and LeRoy and Weimer (1987).

The geologic column (right side of Fig. 4) shows thicknesses of the various rock formations, rock types (e.g., sandstone, basalt, gneiss, etc.), and the approximate time spans represented by the rock formations with unconformities[10] between them. The oldest rock formation is at the bottom of the column, and the youngest is at the top. Descriptions in the center of the column locate the various rock formations in Golden. Adding up the thickness of all the rock formations above the Precambrian gneiss comes to 13,525 feet for the Golden area. That is a lot of rock with a lot of stories.

Organization

Not a geologist? Interested in science? Interested in history? Our book is written for you. Words in blue are defined in the Glossary. Web hyperlinks are underlined in blue. A Geology Primer develops a shared language of concepts and terms. It also describes how to read the geologic map and geologic column shown in Figure 4. Footnotes in the Geology Primer have website hyperlinks to resources where you can dive deeper into a given topic.

Here are a few questions to decide if you want to start with the Primer first. Do you know the types of sedimentary rocks? Do you know what a fault or fold is? Do you understand the concept of Deep Time? How about the Law of Superposition? If you answered "no" or "I'm not sure," to all of these questions, or you would just like a refresher, read through the Geology Primer first. Otherwise, just start into the book. You can always jump to the Primer.

The book consists of two informal, nearly stand-alone parts: geology and mining history. The stories of Golden's bedrock geology are in Chapters 2 through 5. Chapter 6 explains how the Ice Ages and earlier erosion history shaped the present landscape. Chapter 7 discusses surface and groundwater resources of Clear Creek and the Golden Valley. The mining history of Golden includes Chapters 8 through 11. These chapters recount how people used, and still use, Golden's geologic resources. Mining in the Golden area has been active from the mid-1800s through today. It spawned important industries, some of which exist today. Many mining legacies (Chapter 12) are left in the city and surrounding area. Some have created problems for urban development. Others became historic sites, parks, and open space. Yet others continue to cause controversy.

---

[10] Unconformities represent gaps between two or more rock formations that may represent geologically long time-spans.

Chapter 2. Primordial Beginnings

The oldest rocks on the west side of Golden, Precambrian metamorphic rock and three sedimentary rock-formations (Fig. 5), tell us about the first two mountain-building episodes in Golden. Precambrian gneiss forms today's Front Range mountains. The Fountain, Lyons, and Lykins sedimentary rock-formations are present at the base of the mountains on the westernmost side of Golden and trend northwest-southeast. The Fountain Formation extends completely across the west side of the city, almost disappearing at the mouth of Clear Creek Canyon. The Lyons and Lykins formations are only present in north and south Golden.

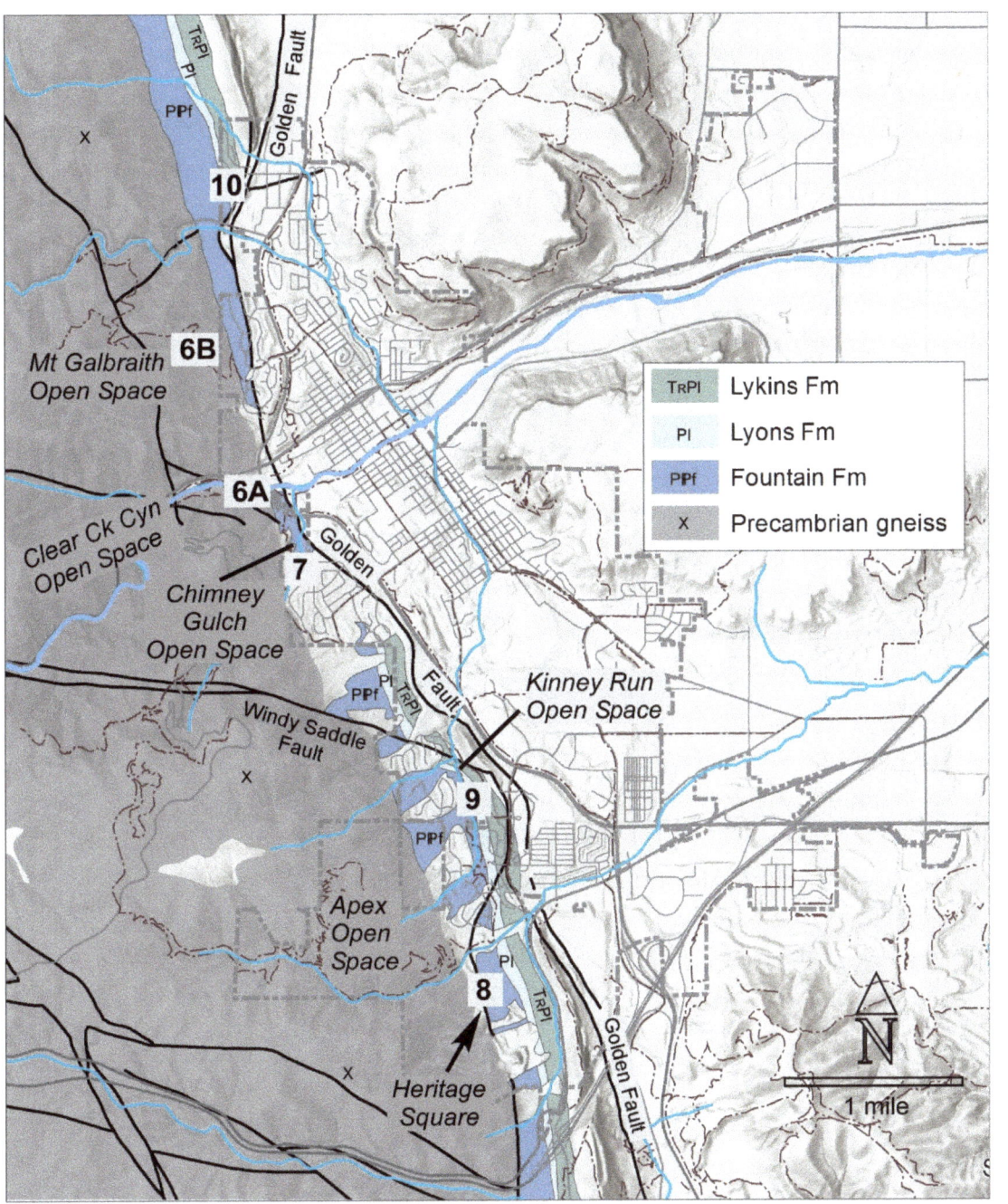

Fig. 5. Geologic map of Precambrian gneiss and the Fountain, Lyons, and Lykins rock formations across Golden. Numbers correspond to Figure locations.

Front Range Metamorphic Bedrock

Metamorphic rocks form the Front Range bedrock in the Golden area, mostly composed of gneiss and schist. Less common rocks are the igneous rock called granite, and veins of quartz mixed with pink feldspar, called pegmatites. All of these rocks were intensely deformed by partial to wholesale

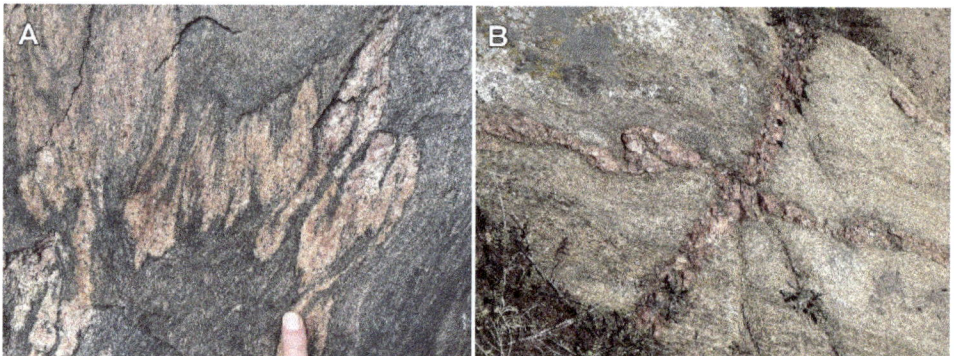

Fig. 6. A) Squiggly texture (from folding) of pegmatite vein in gneiss near the mouth of Clear Creek Canyon. B) Crossing pegmatite veins in gneiss at Mt. Galbraith Open Space Park.

melting, folding, and faulting from massive Earth, or tectonic, forces that gave them a straight to squiggly, banded appearance (Fig. 6A). You can observe all of these rock types along trails in the Open Space Parks of Mt. Galbraith, Clear Creek Canyon, Chimney Gulch, and Apex. At Mount Galbraith Park, spectacular examples of pegmatite veins (Fig. 6B) are visible along the Bluebird Trail. Gneiss and schist represent ancient sedimentary rocks deposited and buried in a series of ocean basins and then metamorphosed under massive pressure deep in the Earth during the first of three mountain-building episodes in Golden around 1700 million years ago (in the Precambrian Era[11]). Pegmatite dikes intruded the gneisses in a later event, near the end of that first mountain-building episode.

Being a relatively difficult rock to erode, gneiss forms the high-elevation ridgeline and uplands in the Front Range on the west side of Golden. That erosional resistance is also why Clear Creek Canyon is narrow and steep-sided.

The Great Unconformity

The "Great Unconformity" is the contact between Precambrian gneiss and the overlying Fountain Formation (Fig. 7).

Fig. 7. The Great Unconformity (dashed line) in Chimney Gulch Open Space: Pennsylvanian Fountain Formation (tilted down to right) overlies banded Precambrian gneiss. View to north.

Early geologists, like F.V. Hayden, immediately recognized that sedimentary rocks on top of metamorphic gneiss represented a huge change in geology. Why is the Great Unconformity[12] so great? The age difference between the two rock types turned out to be an even bigger clue. An ultra-long time,

---

[11] A web hyperlink to the geologic time scale: https://www.geosociety.org/documents/gsa/timescale/timescl.pdf . Geologic, or Deep, time is discussed in the Geology Primer.

[12] An unconformity in rocks is like having a bunch of pages missing in your favorite book! See "unconformities" in the Geology Primer.

about 1360 Ma,[13] is represented between the ages of the two rock units. Geologists derive that time span from the fact that the base of the Fountain Formation is about 318 Ma old, whereas the underlying gneiss is about 1700 Ma old.

What happened during that "Great," 1360 Ma, time-span? The question simply cannot be answered in Golden, because no rocks are left representing that time span. As suggested by rare limestone pebbles in the base of the overlying Fountain Formation, older sedimentary rocks of Early to Middle Paleozoic ages were likely eroded off a rising mountain range during the Ancestral Rockies (see below) mountain-building episode. Outcrops near Colorado Springs and in the mountains around Vail Pass and in Glenwood Canyon also give us an indication of the older sedimentary rocks that may have been stripped off. Those rocks would have told a tale of shallow tropical to subtropical seas with flourishing, but now extinct, marine life forms.

Golden's "Red Rocks"

Lying on top of Precambrian gneiss, three rock formations, the Fountain, Lyons, and Lykins, form the oldest series of sedimentary rocks in Golden. Outcrops of these formations generally form low hills, but the Lyons sandstone forms cliffs along Kinney Run Gulch in south Golden. Originally deposited flat, tectonic forces pushed up the rocks and tilted them eastward due to the much later (Late Cretaceous-Paleogene) Laramide mountain-building episode (Chapter 5). Geologically young alluvial deposits (Chapter 6) and roads, homes, and businesses commonly cover these outcrops today.

Fountain Formation

The Pennsylvanian-Permian sedimentary rocks of the Fountain Formation formed between about 318 and 285 Ma ago. They are the same rocks as those in Boulder known as the Flatirons; in Morrison at Red Rocks Park and Amphitheater; further south in Roxborough as Roxborough State Park; and in Colorado Springs as Garden of the Gods. Not as spectacular or as resistant to erosion as those neighbors, the best Fountain outcrops in Golden are at the former site of Heritage Square[14] (Fig. 8) and along the lower reaches of Chimney Gulch Trail west of US Hwy-6 (Fig. 5).

Fig. 8. "Bulldozer-manicured" outcrop of Fountain Formation at former Heritage Square in south Golden: rocks tilted down about 50° to right (east). Dark red bands are mudstone. Light red bands are sandstone. View to north. Photo taken with permission from Martin Marietta Materials, Inc.

The Fountain Formation is composed of red coarse-grained sandstone and conglomerate interbedded with brick-red mudstone. Energetic rivers eroded the very ancient Ancestral Rocky Mountains to the west and then deposited the sediments in a basin to the east. The sand grains and conglomerate pebbles have **nearly** identical composition to the Precambrian gneiss bedrock on which they lie. The red rock-color comes from a dominance of pink potassium-feldspar in the sandstone and red hematite (iron oxide) in the mudstone.[15]

Above the Great Unconformity, deposition of the Fountain Formation is associated with the "Ancestral Rockies" orogeny, the second mountain-building episode in Golden. The uplifted ancient

---

[13] Ma means "million years," as explained in the Geology Primer. It comes from the Latin: mega-annum.
[14] At the former Heritage Square development, the outcrop was called "Big Rock Candy Mountain." The owners intended to build a castle on top of the planed-off outcrop in the original development scheme (Richard Gardner, Historian, personal communication, and Gardner, 2015, p. 4).
[15] Hubert, 1960.

mountain range occupied almost the same location as today's Front Range. Hence, it became known as the "ancestor" to the "modern" Rockies. The Ancestral Rockies eroded down over time, and the eroded sediment deposited in the adjacent basin became the Fountain Formation. As Shakespeare said, "The revolution of the times make mountains level."[16]

Lyons Formation

By Late Permian time (296-252 Ma), the Ancestral Rocky Mountains had eroded down to low hills, and a sea had advanced from north to south almost to Lyons, Colorado, 35 miles north of Golden. There, the rocks that we call the Lyons Formation formed part of a large coastal sand-dune system. Around Lyons, much of that rock is quarried for building and decorative stone and is known as Lyons Sandstone.

Fig. 9. The Lyons Formation consists of white sandstone (Ss) and conglomerate deposited in the channel of an ancient river, in south Golden in Kinney Run Open Space. Cambria Kiln at bottom left. View to southeast.

In Golden, however, the Lyons Formation has a river-dominated origin, being composed of pebbly sandstone and conglomerate deposited in channels.[17] Lyons sandstone is white because it contains white quartz instead of the pink feldspar contained in the older and redder Fountain sandstone. White Lyons sandstone forms a prominent ridge in south Golden along Kinney Run Gulch (Fig. 9) and along the west side of Pine Ridge Road in north Golden. Fossils in the Lyons Formation are rare because the depositional environment did not favor preserving bones and plants. Animal tracks found in the Lyons near Boulder suggest that ancient dinosaur-predecessors probably cruised around Golden on a semi-arid coastal plain. Some of the track-containing slabs from the Lyons Formation are on display in front of the Jefferson County courthouse in Golden.[18]

Lykins Formation

The mostly red shales of the Permian to Triassic (252-201 Ma) Lykins Formation overlie the sandstone of the Lyons Formation. Because shale is easily-eroded, Lykins outcrops form valleys that are covered with grasses. A well-exposed outcrop in the cut-slope behind the City of Golden maintenance shops, off Catamount Drive in north Golden (Fig. 10), shows the brick-red shale typical of the Lykins Formation, as well as two white beds of limestone.

---

[16] Henry IV, Part 2, Act 3, Scene 1.
[17] Weimer and Erickson, 1976.
[18] Lockley and Marshall, 2014.

The "Bastard" Limestone

When F.V. Hayden visited Golden in 1869, he examined outcrops while riding horseback. His notes record a layer of "bastard limestone" among the red rocks on the west side of town, in the vicinity of today's CSM survey field and along Kinney Run Gulch.[19] What did he mean by bastard limestone?

Fig. 10. Lykins Formation with the Falcon and Glennon limestone members in north Golden at the City of Golden maintenance shops off Catamount Drive. View to northwest.

As it turns out, Hayden was referring to the Glennon[20] and Falcon limestone members of the Lykins Formation (Fig. 10). Those rocks are composed of calcium carbonate ($CaCO_3$), or limestone. But, the limestone was mixed with a lot of sand grains: an "impure" limestone, possibly leading to Hayden's description as "bastard."[21] The limestone beds are present near the base of the Lykins Formation, a few tens of feet above the Lyons Sandstone. They formed in a unique depositional environment, creating a special type of fossil that combined a lot of sand grains as it formed.

Fig. 11. Crinkly laminated stromatolites curve upward, away from pencil point. Glennon Limestone outcrop at same location as Fig. 10.

Stromatolites, an unusual fossil (Fig. 11), make up the limestone of the Glennon and Falcon members. Stromatolites formed from cyanobacteria (a microbe that is also called blue-green algae) that lived as slimy mats in ponds, lakes and/or shallow oceans, such as existed long-ago in the Golden area.[22] Those ancient water bodies were ultra-salty, inhibiting the growth of non-bacterial organisms and often becoming prone to salt formation. It was not a nice place to lie on the beach for a vacation.

Carbonate blobs and quartz sand-grains got stuck on the slime, eventually covering the bacterial mats. As the bacteria died, only the trapped carbonate/sand grains remained. Layer upon layer built up

---

[19] Hayden, 1869, p. 34.
[20] Also called the "Forelle Limestone" by many geologists.
[21] He might also have meant that finding a limestone amongst dominantly sandstone and shale was very odd.
[22] A more modern name for stromatolite is "microbialite," which acknowledges an origin from microbes (bacteria). Stromatolites form today at Shark Bay, Australia, and the Great Salt Lake of Utah.

over time, creating the stromatolite rock, commonly showing pinstripe, crinkly and curvy laminations (Fig. 11). Buildups grew to nodular and pustular mounds up to several feet high. Good examples of the mounds are rare now because the outcrops around Golden became limestone mines, producing mortar for Goldenites in the late 1800s (Chapter 11).

The Great Dying

The boundary between the Permian and Triassic Periods (also the boundary between the Paleozoic and Mesozoic Eras) 252 Ma ago, represents a massive extinction of life. Known to paleontologists as the "Great Dying," more than 90% of all marine species such as Trilobites (ancient arthropods, like horseshoe crabs) became extinct. Yet, extinction of some animal groups gave an opportunity for surviving groups to expand during the Mesozoic Era. Why did this extinction happen? While still debated, evidence is pointing toward a massive outpouring of volcanic gasses and particles (volcanic ash) from Siberia that poisoned the atmosphere and oceans.[23]

Dr. James Hagadorn, Curator of Geology at the Denver Museum of Nature and Science, has been investigating the Permo-Triassic extinction in the Lykins Formation along the Front Range, including the outcrop in north Golden.[24] Aside from stromatolites, the rarity of fossils in the Lykins has made the study difficult. The arid climate and slimy, salty depositional environment was not favorable to much, if any, animal or plant life.

Geologists have debated exactly where the boundary lies within the Lykins Formation for the last 100 years. It potentially crops out in two places in Golden: one near the city maintenance shops off Catamount Drive in north Golden (Fig. 10), and the other in the Kinney Run Open Space near the intersection of Kinney Run Gulch and Eagleridge Drive in south Golden. The elusive boundary location has never been confirmed at either site.

---

[23] See Fielding et al., 2019, for discussion.
[24] Hagadorn et al., 2016.

Chapter 3: Jurassic Park and the Western Interior Seaway

The next six rock formations, stacked on top of the Lykins Formation (Fig. 4), represent the time of "Jurassic Park" where dinosaurs roamed freely across the landscape, followed by a change to a widespread ocean. The oldest Jurassic (199-145 Ma) rocks are the red and gray-green shale, siltstone, gray limestone, and white sandstone of the Ralston Creek and Morrison formations (Fig. 12). The overlying Early Cretaceous Dakota Group, consists of yellow sandstone and conglomerate, gray-brown

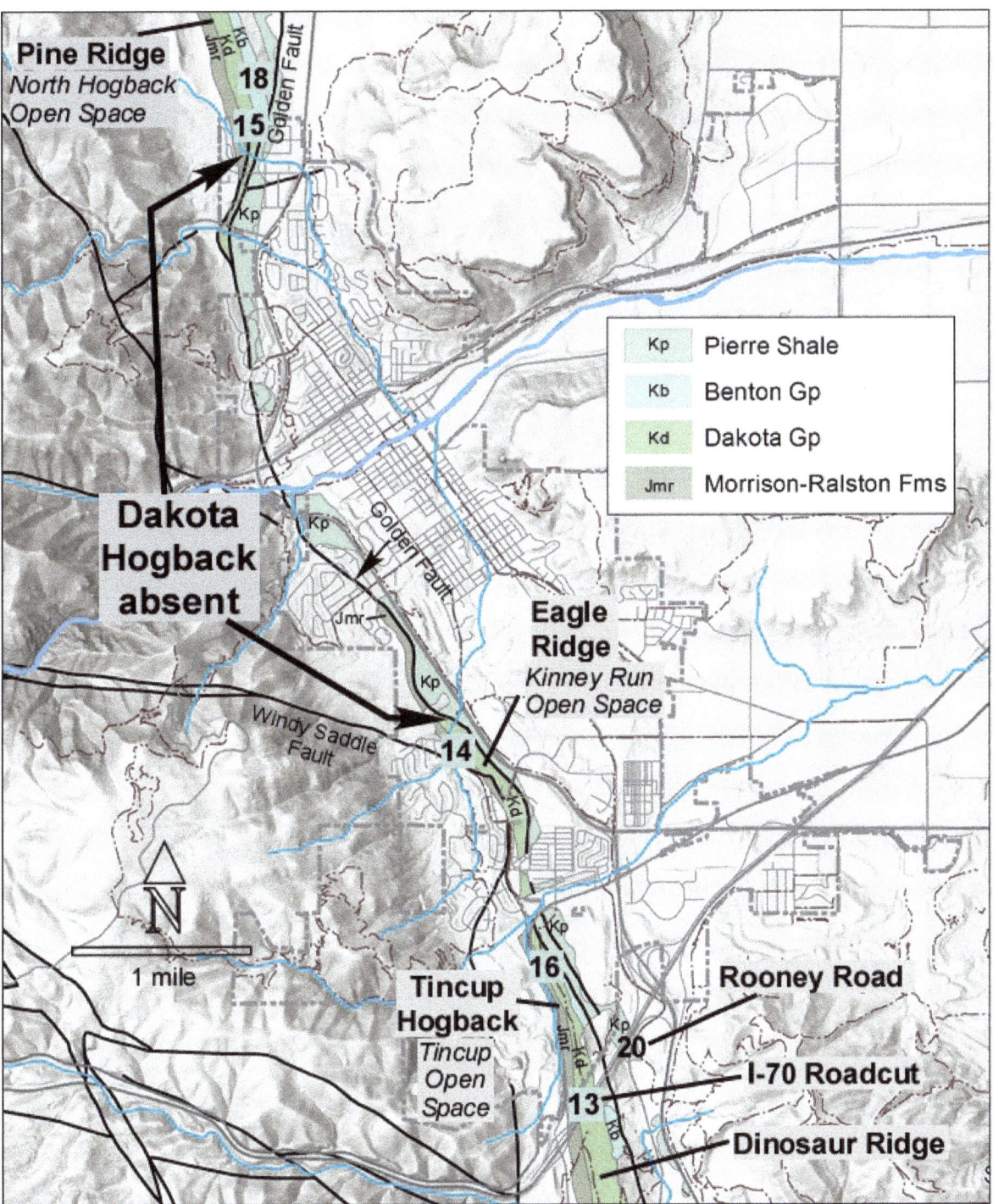

Fig. 12. Geologic map of Jurassic to Late Cretaceous rocks, showing local names for the Dakota Hogback. Numbers correspond to Figure locations.

and red siltstone, and black shale. On top of the Dakota Group are two rock formations that barely crop out in and around Golden: the early part (100 to 70 Ma) of the Late Cretaceous Benton Group through the Pierre Shale. A third, the Niobrara Formation, between the Benton and Pierre Shale, is completely buried below the land surface in Golden. Together these last three rock formations represent a time when Golden was completely under a huge ocean, known as the Western Interior Seaway.[25]

Rocks of the Ralston and Morrison formations and the Dakota Group form a prominent ridgeline called the Dakota Hogback, extending from Colorado Springs to Fort Collins. In the Golden area, the Dakota Hogback is given different names where natural and human-made features break it into sections (Fig. 12). South of the I-70 roadcut it is called Dinosaur Ridge (Fig. 13). North of the roadcut, it is called Tincup Ridge, which ends on the south side of U.S. 40/Colfax at Lena Gulch. Across the Golden Ridge

Fig. 13. South side of the I-70 roadcut (Dinosaur Ridge side) shows tilted rocks of the Morrison and Ralston Creek formations and Dakota Group. The roadcut is a designated Point of Geological Interest in the national Interstate Highway system. View to south.

neighborhood, the Hogback forms low hills and re-emerges as a prominent ridge, called Eagle Ridge, southwest of the intersection of Heritage Road and US Hwy-6. Absent through central Golden, the Hogback reappears in north Golden where it is known as Pine Ridge, or simply the North Hogback.

Golden's Jurassic Park

The popular book and movie, "Jurassic Park," centered on the re-creation of extinct dinosaurs that roamed the Earth throughout the Mesozoic Era (Age of the Dinosaurs) from 252 to 66 Ma ago. Glossing over the fact that T. rex did NOT live during the Jurassic Period, other types of dinosaurs, such as Apatosaurus (sometimes also called Brontosaurus), Stegosaurus (the Colorado state fossil), and Allosaurus, a 2-ton predator, thrived in Golden during the Jurassic Period.

The first discovery of dinosaur bones in Colorado was along the west side of Dinosaur Ridge in the Dakota Hogback (today's Dinosaur Ridge) near Morrison. Arthur Lakes discovered the bones in 1877 and sent them off to Professor Othniel Charles Marsh at the Yale Peabody Museum in Connecticut.[26] While beginning a productive association, the correspondence also promoted a famous feud, known as the "Bone Wars," between Marsh and his arch-rival Edward Drinker Cope of the Academy of Natural Sciences in Philadelphia. Much later, geologists recognized brontosaur tracks in Morrison "pond" deposits: the so-called "Bronto-bulges" along the west side of Alameda Parkway at Dinosaur Ridge. The Morrison dinosaur quarries of Arthur Lakes and the tracks are now part of Morrison-Golden Fossil Areas NNL,[27] managed by the Friends of Dinosaur Ridge. Alas, the Morrison Formation in Golden has no reported bones or dinosaur tracks. Perhaps someone will find some, someday!

---

[25] The Western Interior Seaway reached from the Arctic Ocean, south to the Gulf of Mexico, effectively splitting the United State and Canada into two separate land masses. On the west it reached into Central Utah, and on the east, it covered Kansas. Ocean depths ranged from 100 to up to 1000 feet at various times.

[26] For a complete history of Lakes' quarries see Simmons and Ghist, 2014.

[27] NNL stands for National Natural Landmark, the first level of National Park Service recognition for an area of scientific and/or scenic interest.

Subtle outcrops of the Morrison and its basal part, the Ralston Creek Member,[28] consist of vegetation-covered red and gray-green siltstone, limestone, and shale with scattered white sandstone (Fig. 14). Lens-shaped sandstone beds represent ancient river deposits. Gray-green shale (e.g., Fig. 13) represents ancient lake deposits. All of these rock types fit with an ancient landscape across which dinosaurs roamed freely: a huge subtropical inland plain crossed by rivers and dotted with lakes.

Fig. 14. The Morrison and Ralston Creek formations on west side of Eagle Ridge form the reddish slopes below ridge-top of yellow Dakota sandstone on the skyline. Small arrows point to sandstone beds deposited by rivers. View to east.

During Morrison deposition, a rising volcanic mountain chain along the West Coast of North America spewed volcanic ash across the entire Western U.S. At the same time the Western Interior Seaway began to encroach into Montana and Wyoming from the far northern arctic.[29] The end of Morrison deposition was a time of landscape stability followed by erosion for a geologically long time, creating a major unconformity. Hence, the contact between the Morrison Formation and overlying Dakota Group represents about a 35 Ma time gap.[30] Nothing is left to tell us exactly what happened geologically during that time span in Golden and much of the rest of Colorado.

Shorelines of the Western Interior Seaway

Around 110 Ma ago, in the Early Cretaceous Period (145-100 Ma), the Western Interior Seaway kept advancing from the north and arrived in the Golden area. The Dakota Group represents that oceanic return and includes the South Platte and underlying Lytle formations.[31] Rocks of the Dakota Group formed within a network of ancient valleys carved into the underlying Morrison Formation. Eventually the ancient valleys filled up, and became overtopped and buried as the depositional environment changed from land to a large shallow bay, or estuary. Extensive tidal flats with ripple-marked sandstone (Fig. 15) were common. Brackish

Fig. 15. Wave-ripple marks on several steeply tilted sandstone beds in the upper Dakota Group along Pine Ridge Hogback in north Golden Hogback Open Space. View to west.

---

[28] Renamed the Ralston Creek Member of the Morrison Formation by Carpenter and Lindsey (2019, p.8), it still appears on geologic maps as the Ralston Creek Formation.
[29] Geologists may want to know that this change was caused by the start of subduction on the West Coast. See Geology Primer.
[30] As per the laws of physics, gaps in time are impossible, but the phrase is commonly used to mean a gap in the rocks that represents a long time: also explained in the Geology Primer.
[31] In the subsurface (underground) of the Denver Basin, the South Platte Formation consists of the "J Sandstone" at the top, underlain by the Skull Creek Shale, and lastly, the Plainview Sandstone, all of which overlie the Lytle Formation (LeRoy and Weimer, 1971).

water in the estuary had an acidic, organic-rich chemistry[32] that led to forming deposits of the clay-mineral kaolinite, a pure aluminum-silicate clay called fire-clay. The kaolinite-rich estuary deposits alternated with ripple-marked sandstone of the tidal flats as sea level fell and rose periodically. Thus, in most of Golden, the uppermost part of the South Platte Formation consists of visually striking, stark-white sandstone slabs along the east side of the Hogback.[33] The white slabs are the remnants of clay mines that left behind deep trenches after mining out fire-clay seams in the early 1900s (Chapter 10).

Crawling with Ancient Life

Golden's Early Cretaceous subtropical ecosystem was rich with life. Trees flourished near lagoons, leaving behind impressions of their trunks where they fell and became preserved. Fallen leaves left impressions in the claystone, as testimony to the abundant subtropical vegetation. Worms and other invertebrates lived in the soft sediment, leaving behind burrows that are visible in the rocks today. Prehistoric birds flew across the sky, and primitive small mammals scurried through the vegetation undergrowth. The environment was also ideal for dinosaurs of the time (again, not T. rex), such as hadrosaur foragers, and the predator Acrocanthosaurus, known from their tracks left in wet sand. Crocodiles, which swam through tidal channels, left claw marks on the soft sandy bottom (Fig. 16). Birds also left tracks as they walked across tidal flats foraging for food.

Fig. 16. Crocodile swim marks in upper Dakota Group along the Tincup portion of the Dakota Hogback in south Golden.

World-renown dinosaur trackways in the South Platte Formation are spectacularly preserved and protected on the east side of Dinosaur Ridge near the town of Morrison where the Friends of Dinosaur Ridge conducts tours and programs. In south Golden, dinosaur tracks, crocodile swim marks, and fossil leaves are reported on Tincup Ridge.[34] In north Golden, heron-like Ignotornis,[35] crocodile, and dinosaur tracks are found at the Early Bird site[36] which is a part of the Morrison-Golden Fossil Areas NNL and protected as part of the 18-acre North Hogback Open Space area in the City of Golden.

Underwater for a Long Time

By 100 Ma the Western Interior Seaway flooded across the top of the Dakota Group rocks and covered most of Colorado, Kansas, Nebraska, North and South Dakota, New Mexico, eastern Utah and eastern Wyoming. It was a huge seaway. Golden stayed underwater for the next 30 Ma. The shale rocks representing the first 20 Ma, between 100 and 80 Ma ago, are those of the Benton Group and the Niobrara Formation. The Pierre (pronounced "peer") Shale represents the last ten Ma, from 80 to 70 Ma. Due to the Golden Fault System, though, most of these oceanic shale deposits are deeply buried in the subsurface (Chapter 5).

---

[32] Waage, 1961.
[33] The white rocks are part of the J-Sandstone member. The Plainview and Skull Creek members are not present across most of Tincup Ridge and all of Eagle and Pine Ridges.
[34] The Hawks Nest site of Lockley et al., 2014.
[35] Lockley et al., 2009, and Houck et al., 2010.
[36] Ibid.

How deep was the ocean that covered Golden? During deposition of the Benton Group, Niobrara Formation, and most of the Pierrre Shale, water depth probably ranged from 300 to 1000 feet. To safely explore the bottom, you would have needed a submarine. But, by the end of Pierre deposition, the ocean was getting to snorkeling depth, about 30 to 60 feet. As sea level continued dropping, the Western Interior Seaway retreated to the south and east, eventually leaving for good.

Benton and Niobrara Ocean Worlds

The Cretaceous ocean teemed with life in the water and on the seafloor. Formidable predators like Mosasaurus and Plesiosaurus, along with giant sharks, swam in the ocean. Ammonites (relatives of the chambered Nautilus and squids), some reaching up to three feet in diameter, floated and jetted through the water, along with numerous fish. During deposition of much of the Benton and all of the Niobrara formations, tiny algae (Fig. 17) were super-abundant in the upper parts of the water where sunlight could penetrate. When they died, the tiny calcium-carbonate ($CaCO_3$) algal shells fell to the ocean floor, forming extensive chalk beds that covered all of eastern Colorado, including Golden, and western Kansas. Clams, worms, and snails also lived on the seafloor where there was enough oxygen to sustain them. Often during Benton and Niobrara deposition, though, the seafloor area was dysoxic, meaning it had a very low oxygen content, thus inhibiting seafloor life.

Fig. 17. Ultra-tiny (scale bar is ten microns) fossils of an algae called a coccolith make up the chalks of the Niobrara Formation. Image from the USGS Core Research Center, well-number D967.

Only about the bottom 30 feet of the 810-foot thick Benton Group rocks are visible in Golden.[37] These subtle and rare outcrops lie immediately above the white sandstones of the Dakota Group on the east side of the Dakota Hogbacks in north (Fig. 18) and south Golden. The remaining Benton Group rocks are present far below the ground surface. Similarly, the overlying shale formation, the 330-foot-thick Niobrara, is nowhere exposed in Golden, because it, too, lies far below the ground surface. Thus, 760 feet of these shale rocks are not exposed at the surface in Golden.

Pierre Ocean World

During Pierre deposition, the Western Interior Seaway also teemed with life, but the silt-filled water made for a poorer environment for microscopic algae than did the clearer and cleaner ocean water of the Niobrara environment. Swimming in the water were similar animals as before, such as straight-shelled ammonites (Baculites) and giant sharks. However, more commonly than in the

Fig. 18. A lone outcrop of the basal Benton Group (B) forms the mound against the white Dakota sandstone in north Golden on the east side of Pine Ridge. View to southwest.

---

[37] The Mowry Shale forms the base of the Benton Group.

Niobrara ocean, snails, clams (Fig. 19), and worms flourished on the Pierre sea floor, especially on submarine sand-bars scattered across ocean floor. In other parts of Colorado, these ancient offshore sandstone layers supported abundant clam "meadows" that are now popular with collectors, because fossil clams are abundant and easy to find.

Fig. 19. Mold of the clam *Inoceramus* from near Kremmling, CO that was common in nearshore areas of the Western Interior Seaway.

Notably, the Pierre Shale is about 8000 feet, or about 1.5 miles, thick. That is 8-times thicker than the 1000-foot thick Fountain Formation! Pile 8000 feet of Pierre over all of Colorado, and it is a huge volume of rock! All of those particles came from a giant mountain range 450 miles to the west, known as the Wyoming-Utah Thrust Belt. That rising and eroding mountain range sent gobs of clay and silt into the Western Interior Seaway. But today, only the upper 300 feet of the Pierre Shale crops out in Golden, mostly under US Hwy-6. Like the Benton and Niobrara formations, rocks of most of the Pierre Shale are present far below the ground surface due to the Golden Fault System (Chapter 5).

The Pierre Shale in the Golden area is not really just shale. Rather it is dominantly siltstone[38] and claystone with lesser amounts of sandstone, usually in thin beds (Fig. 20). In the Pierre Shale, silt grains are mixed with tiny clay minerals. Unfortunately, for homeowners, commercial builders, and highway departments, the Pierre Shale contains sticky clay minerals (smectite) that ruin building foundations and roads by swelling when wet. Standard advice among field geologists is to never drive on the Pierre Shale when it is wet, because the sticky clay will make your car, and any tow truck you might call, become totally stuck in the muck. You will not leave until the ground dries.

Fig. 20. The upper part of the Pierre Shale along Rooney Road south of I-70 tilts steeply down to the east (left). Thin sandstone fins stick up among more easily eroded siltstone and claystone. View to south.

---

[38] Silt grains are the diameter of a hair and mostly invisible to your eye. Clay-size grains and minerals are less than ten microns, totally invisible to the eye.

Chapter 4: Good-bye Seaway, Hello Mountains

Next in the geologic progression (Fig. 4) are five five rock formations, the latest (70-66 Ma) Late Cretaceous Fox Hills, Laramie, Arapahoe, Denver, and Green Mountain formations. They are the true rock stars of Golden's geologic history. They tell the story of the permanent exit of the Western Interior Seaway and the building of the Rocky Mountains that we see today. That mountain-building episode is named after the Laramie Formation: the Laramide Orogeny (Chapter 5). Golden rocks have the best remaining geologic evidence for the early part of this episode in northeastern Colorado.

The Fox Hills and Laramie formations extend northwest to southeast across the west side of Golden (Fig. 21). These rocks consist of sandstone, siltstone, claystone, and coal that were deposited first at the shoreline, and later within a large delta plain covered with swamps. After Laramie deposition (69-68 Ma), the ocean never again covered the Golden-Denver area. The Arapahoe, Denver, and Green

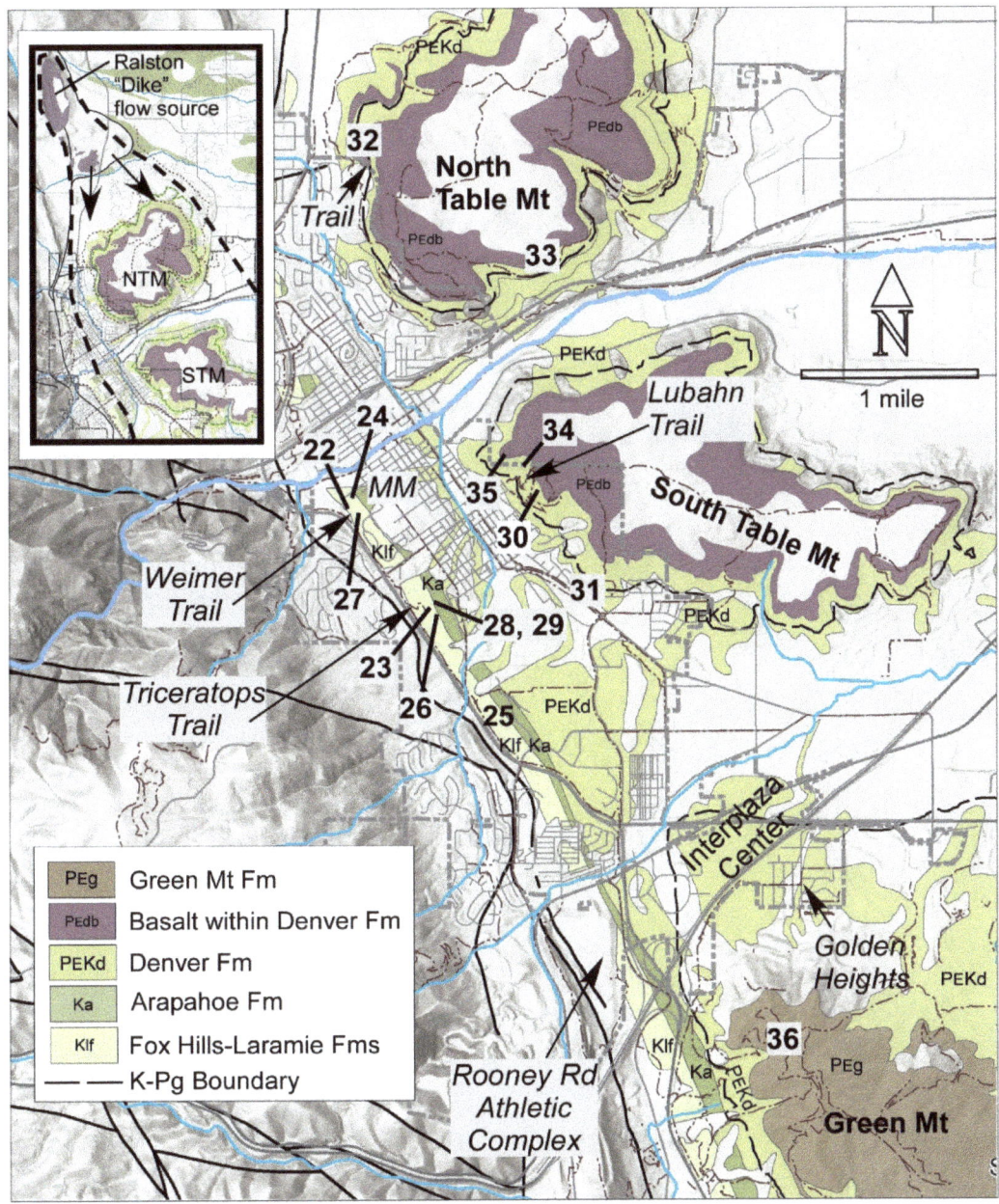

Fig. 21. Geologic map of the Fox Hills, Laramie, Denver (including basalt flows), and Green Mountain formations across Golden. Inset map shows source of basalt flows from Ralston Dike, two miles northeast of North Table Mountain. "MM" is the Mines Museum of Earth Science. Numbers correspond to Figure locations.

Mountain formations are present on the central, east and south sides of Golden (Fig. 21). Those formations are composed of conglomerate and coarse sandstone deposited by energetic rivers that flowed out through the rising mountains near Golden's doorstep. Large amounts of siltstone were deposited across the floodplains next to the river channels. Several lava flows in the Denver Formation poured out over the land surface 66 to 65 Ma ago, filling river channels. Known as the Table Mountain basalt flows, they now form the caprocks on the two Table Mountains. Just to add interest, an asteroid hit the Earth's surface a short time after the oldest lava-flow, ending the Age of Dinosaurs. As geologists say, "It was an exciting time in Golden's geologic history!"

Beaches and Swamps

It is hard to imagine that the area now occupied by Fossil Trace Golf Course used to be a beach and then the swamps and bayous of a large delta system. The resulting rock formations have been mined for clay and coal resources, and paleontologists have studied the famous plant and animal (dinosaur and mammal) fossils. The Laramie Formation, in particular, includes two parts of the Morrison-Golden Fossil Areas NNL and associated walking trails because of its nationally recognized fossil importance.

Fox Hills Beach

As the Western Interior Seaway retreated eastward, the ocean in Golden got shallower, eventually becoming a shoreline with a beach, which is represented by the Fox Hills Formation. The Fox Hills interval consists of two or three sandstone layers separated by shale. Geologists agree that each sandstone layer, also called a "tongue," records a minor advance and retreat of a shoreline before finally becoming the beach at the topmost layer. The "beach" sandstones are eye-catching because they are almost pure white (Fig. 22), being composed of quartz sand. In the middle 1800s during Golden's early history, the Fox Hills Formation was mined for a short time as a source of pure quartz (silicon dioxide, SiO_2) for glass making.

Fig. 22. White "beach" sandstone of the Fox Hills at Stop 6 on Weimer Geology Trail at CSM. Pierre Shale is to right. All beds are overturned, tilted down to the right, or west. View to south.

Fox Hills outcrops are rare around Golden, having been covered by roads and trails along US Hwy-6 and beneath the former Jefferson County landfill at the the Rooney Road Athletic Complex in south Golden. Aside from the Rooney Road outcrop south of I-70 (Fig. 20), the best Fox Hills outcrop is found along the Weimer Geology Trail on the CSM campus (Fig. 22), where it is upside down (overturned) due to folding and faulting. With its many sandstone tongues reaching far down below the ground surface and across the Denver Basin to the east, the Fox Hills Formation is also a groundwater-bearing unit, or aquifer (Chapter 7).

The Swamp-Infested Laramie Formation

As the sea retreated farther and farther eastward toward Kansas, Golden became covered by the swamps of a delta plain, represented by rocks of the Laramie Formation. Such rocks consist of thick sandstone and siltstone beds alternating with coal and claystone beds, each ranging from 3 to 20-feet thick. A major coal bed (also called a "seam"), 8-14 feet thick, extended north from the CSM campus to Ralston Creek, a distance of about seven miles. Known as the "White Ash" coal seam, it was heavily mined from the late 1860's to the early 1900's (Chapter 9).

Studied extensively by Dr. Robert Weimer,[39] Laramie sandstone beds originally formed as crevasse splays across the broad delta plain. Relatively flat and wedge-shaped, these beds formed low mounds, or sand bars, above an otherwise watery environment of shallow lakes full of water-loving vegetation: swamps. Fine-grained clay and silt periodically filled the lakes when floods inundated the delta plain. Abundant subtropical vegetation, consisting of palm (Fig. 23), magnolia, and sycamore trees as well as ferns and grasses flourished in and around the swamps. As an analog, think of the extensive swamps of the modern Mississippi River delta in southern Louisiana: full of trees and other plants living in the water and teeming with animals like crocodiles, fish, and worms. Indeed, the Laramie swamps also supported abundant animal life, as shown by dinosaur, reptile, mammal, and bird tracks as well as insect and worm tracks, now preserved at the Triceratops and Weimer geology trails, part of the Morrison-Golden NNL.

Fig. 23. Palm-frond impressions along Triceratops Trail.

Dead plants accumulated in the swamps where decay was very slow due to acidic waters. Eventually sand and mud from river floods filled in the swamps, and the decaying plant material slowly became buried, forming peat. As layer upon layer built up, the increasing pressure from overlying layers changed the peat to coal, which is composed of carbon, sulfur, iron, methane, carbon dioxide, water, and other impurities.

Clay is interbedded with coal in the Laramie Formation, which explains why historic clay and coal mines are in similar locations. Laramie clays have more impurities than those in the Dakota Group, consisting of alternating white and red clay beds (Fig. 24). Because they also were more abundant, Laramie clay beds were mined for a longer time period. In any case, Laramie clay beds formed a large part of Golden's mining history from the 1870s and through today, fostering a thriving brick industry in Golden until 1963 (Chapter 10).

The now-upright sandstone fins of the Laramie Formation form part of the tapestry of Golden's

Fig. 24. Vertically tilted red and white claystone beds in the Laramie Formation on the Weimer Geology Trail at the west end of 12th Street. White beds contain kaolinite. Reddish beds contain iron. View to south.

---

[39] Dr. Robert J. Weimer (1925-2021) was Professor of Geology at CSM from 1957 to 1983. His studies of the Laramie, Dakota, and other formations in Golden and Colorado were ground-breaking for geoscience.

Fig. 25. "Fin" of white Laramie sandstone on the west side of the Jefferson County Laramie Building. View to northeast.

landscape. They are prominent features on Fossil Trace Golf Course and along the east side of US Hwy-6 between Heritage Road and State Hwy-58. The fins also endured when Jefferson County built some of its county buildings. As one story goes, the high sandstone outcropping next to the Jefferson County Laramie Building (Fig. 25) was saved at the request of the Golden Planning Commission. One might say that the rock fin won out over the expectation for a window view. And, of course, the Laramie Building is well named.

Triceratops and Bob Weimer Mines Geology Trails

Triceratops Trail (Fig. 21) on the east side of Fossil Trace Golf Course was established in 2004 through the efforts of "T" Caneer, a longtime volunteer with the Friends of Dinosaur Ridge. It became part of the Morrison-Golden NNL in 2011. Accessed from the regional trail along US Hwy-6, Triceratops Trail is a path through time in the trenches of the former Rockwell Mine operated by the Parfet family from the late 1800s to 2002 (Chapter 10). Several displays show spectacular dinosaur, bird, mammal, and insect tracks, along with palm fronds and other leaf impressions. One of the most famous tracks is a probable T. rex toe-print among the nearby Edmontosaurus dinosaur tracks. Although difficult to absolutely prove, if true, it would be the second one confirmed in the U.S. Famous Triceratops tracks (Fig. 26), some of the first ever reported in the world, are prominent in the outcrops. Last but not least, other exhibits show how clay was mined between the vertically tilted sandstone beds. And, the view across Golden from the top of the Trail is splendid.

Fig. 26. Natural cast of a Triceratops footprint at Triceratops Trail. View is the bottom of the dinosaur track, and the toes are pointed left. View to east.

The Bob Weimer Mines Geology Trail (Fig. 21) at CSM was made part of the Morrison-Golden Fossil Area NNL in 2011 and dedicated to Dr. Robert (Bob) J. Weimer who was instrumental in its establishment. The trail begins and ends at the Mines Museum of Earth Science (Fig. 21). The Weimer Geology Trail is a walk through the Pierre, Fox Hills, Laramie, and Arapahoe formations, entirely on the CSM campus. Eight stops along the trail have interpretative signage explaining geology and mining history. The upper part of the Weimer Geology Trail crosses the former Rubey Mine operated by the Parfet family from 1877 until the area became part of the CSM campus in 1964.

Fig. 27. Fault plane with vertical grooves (arrows) made from grinding action of a fault during movement. Upper part of the Weimer Geology Trail. View to southwest.

Geologic features on the upper trail include spectacular dinosaur tracks of ceratopsians, hadrosaurs and predatory therapods scattered among palm fronds and tree-logs. One outcrop has abundant smooth fault planes (Fig. 27) with grooves, called slickensides, that formed by grinding when a fault moved. The lower part of the trail, near the football and soccer fields, has an exhibit at the surface location of the entrance to the White Ash Coal Mine, where a memorial honors the ten miners who drowned in the White Ash Coal Mine tragedy in 1889 (Chapter 9).

Mines Museum of Earth Science

Before walking along the Weimer Geology Trail, take time to visit the Mines (CSM) Museum of Earth Science at 1310 Maple Street ("MM" on Fig. 21), where you can pick up a guide to the geology and paleontology along the trail. The museum began with the collections of Arthur Lakes in 1874. It became formally housed in Berthoud Hall (the Geology building at 17th and Illinois Street) in 1940 and moved to its current location in 2003.[40] The museum has a huge collection of minerals, many from Colorado's Rocky Mountain mining districts. Some of Arthur Lakes'[41] fossil collection is housed there, as well as some of the minerals he collected in mining districts around Leadville, Colorado. A large block of the Colorado State rock, the white Yule Marble, lies outside the front entrance. Inside is a spectacular specimen of Colorado's State mineral, rhodochrosite. The Museum is well worth a visit.

Volcanoes and Rushing Rivers: The Rockies at Golden's Doorstep

Lying on top of the Laramie Formation (Fig. 4), the rocks of the Arapahoe, Denver, and Green Mountain formations tell the story of a huge change at Golden's doorstep. The lazy rivers and bayous of the Laramie Formation were never to be seen again.

A Rising Mountain Range: The Arapahoe Formation

The Arapahoe Formation[42] consists of conglomerate and pebbly sandstone that lies sharply next to (and on top of) the Laramie Formation (Fig. 28). Such a dramatic change in rock types represents a large change in depositional environment. Conglomerate pebbles and cobbles eroded from a rising mountain range represent much more energetic river deposition than the lazy bayous of the Laramie Formation. Geologists think that the Arapahoe Formation likely spread out across the landscape as alluvial fans[43] in distinct contrast to the swamps of the Laramie Formation. Conglomerate clasts (pebbles and cobbles) are made of a variety of rocks:

Fig. 28. Vertically tilted beds at the base (arrow) of the Arapahoe Formation conglomerate along the 15th fairway at Fossil Trace Golf Course. View to northeast.

---

[40] Eckley, 2004, p. 100.
[41] Arthur W. Lakes (1844-1917) is considered by some to be the "Father of Colorado Geology." He is famous for his dinosaur quarries and discoveries along Dinosaur Ridge in 1877 (Chapter 3) as well as being professor of Geology at CSM from 1880 to 1893. Lakes also collected fossils on Green Mountain and South Table Mountain, finding the first known Tyrannosaurus rex tooth in North America.
[42] Technically, conglomerates of the so-called Arapahoe Formation are the base of the "D1 sequence" of Raynolds (2002), which includes the entire Denver and Green Mountain formations. We use the nomenclature of Kellogg et al. (2008) to match their geologic map units.
[43] Raynolds and Johnson, 2003, p. 172-173.

chert, quartzite, granite, gneiss, and fine-grained extrusive volcanic rocks (Fig. 29). Igneous and metamorphic clasts came from the Precambrian basement rocks of the Front Range. Distinctive fine-grained extrusive volcanic pebbles were eroded from volcanoes that had formed west of Golden, probably near Granby, Colorado.[44]

The sharp change in rock type and depositional environment is a type of unconformity. But in this case, the change does not represent nearly as much geologic time as the Great Unconformity between the Fountain Formation and underlying Precambrian metamorphic rocks (Chapter 2). In fact, the Arapahoe Formation is only about 1 to ½ Ma younger than the Laramie Formation: a small difference from a geologic perspective.

Fig. 29. Arapahoe Formation cobbles. Extrusive volcanic cobble (left) with tiny white crystals, and intrusive granite cobble (right) with large crystals. Scale in inches. From the outcrop at Fossil Trace Golf Course shown in Fig. 28.

What do the conglomerates of the Arapahoe Formation tell us besides a change to energetic rivers? First, they tell us that the Western Interior Seaway had already retreated completely away from Golden, never to return. Second, they tell us that a mountain-building episode, or orogeny, was happening somewhere west of Golden. As noted earlier, that episode came to be known as the Laramide Orogeny.

Denver Formation: Mass Extinction and Lava Flows

Forming the eastern skyline of Golden, North and South Table Mountains are composed of the Denver Formation, which consists of sedimentary and volcanic rocks. The sedimentary rocks consist of alternating conglomerate, sandstone, and siltstone: the whole spectrum of clastic sedimentary rock types. The volcanic rocks consist of four[45] separate basalt-lava flows, two of which cap the tops of North and South Table Mountains.

Ancient river channels filled with conglomerate are scattered throughout the Denver Formation below the lava flows capping the North and South Table Mountain. The channel-filling conglomerates form lens-shaped outcrops, mimicking a channel cross section, such as that seen from the Lubahn Trail near Castle Rock (Fig. 30). Other ancient channels are visible in the high roadcuts along State Hwy-58 on North Table Mountain. Composition of pebbles in the ancient river channels is totally unlike that of the Table Mountain basalt flows. Rather, the pebbles are commonly andesitic and derived from the actively eroding volcanoes near Granby, Colorado, like the clasts in the Arapahoe Formation.

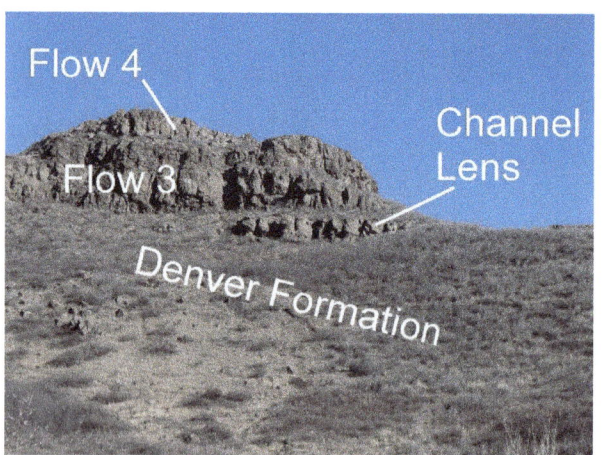

Fig. 30. Lens-shaped river-channel deposit in the Denver Formation below the basalt lava-flows capping South Table Mountain. View to southeast from Lubahn Trail.

---

[44] See Cole and Braddock, 2009, and Dechesne et al., 2011, Plate 13, for discussions of Late Cretaceous volcanism in north-central Colorado.

[45] Originally only three flows were recognized by Glenn Scott (1972) and Richard Van Horn (1972), but Harald Drewes (pronounced like "Davis" with an "r") recognized four flows based on new mapping in the early 2000s (Drewes, 2008), a fact that only reinforces how new things can be found with new, careful work.

The Denver Formation, particularly on South Table Mountain, has been a rich source of plant and animal (mammal and dinosaur) fossils since the 1860s with collections by Edward L. Berthoud[46] and Arthur W. Lakes, among others. Plant and animal fossils found at the base of the Denver Formation were similar to those found in the underlying Arapahoe and Laramie Formations. T. rex and Triceratops still roamed around on the newly-energized river floodplains. But fossils found higher up (younger) in the Denver Formation showed a dramatic change in animal and plant life.

K-Pg Boundary: Another Mass Extinction

The Cretaceous-Paleogene (K-Pg, formerly K-T, Cretaceous-Tertiary) boundary is the time in the Denver Formation when the dinosaurs became extinct. It is precisely age-dated at 66.052 ± 0.043 Ma.[47] As we now know, the boundary is marked by a unique zone of deposits just 5 inches thick in Colorado. The zone includes a unique layer enriched in the element iridium. Even though disappearance of dinosaurs, in particular, initially defined the K-Pg boundary, it took many years to figure out that the ultimate cause of extinction was an asteroid that struck the Earth's surface at the Yucatan Peninsula in Mexico, wreaking havoc across the planet. Massive dust and forest fires caused by that impact obscured the Sun, cooled Earth's atmosphere, and destroyed vegetation, all of which brought dinosaurs to extinction, as recounted by Walter Alvarez[48] in his popular book, "T. Rex and the Crater of Doom."

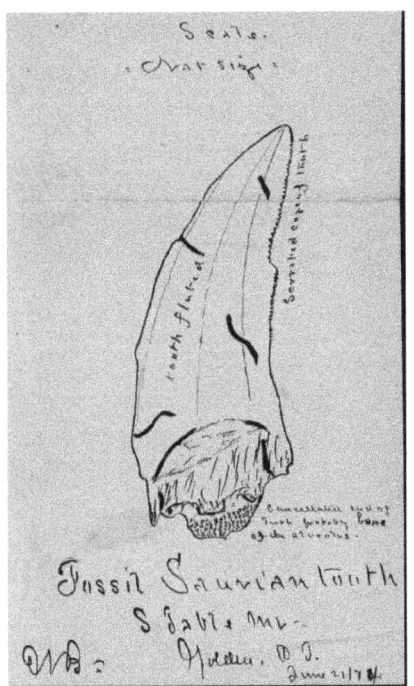

In 1874, Peter Dotson, a student, and Arthur Lakes collected a "splendid saurian tooth" at the base of the southwest flank of South Table Mountain in Golden. Edward L. Berthoud sent a sketch of that tooth (Fig. 31), and later the tooth itself, to O.C. Marsh[49] at the Yale Peabody Museum where it lay forgotten and unidentified for over 100 years. The tooth is now recognized as the first discovery of T. rex in North America.[50] Also, in the 1870s, the upper flanks of South Table Mountain yielded a lot of bones that were at first misidentified as ceratopsian dinosaurs (Triceratops). Later, the bones turned out to be those of mammals. Because of the mis-identification, though, the exact position of the dinosaur-to-mammal change across the K-Pg boundary was not recognized until the late 1930s and early 1940s.

South Table Mountain was the first place in North America where the K-Pg (aka K-T) boundary was confirmed. In 1943, Roland W. Brown[51] published a paper in which he described finding Cretaceous Triceratops bones below a zone on the southeast side of

Fig. 31. Berthoud's note showing the first T. rex tooth found in North America. His initials are at bottom left, but Arthur Lakes made the sketch. Yale Peabody Museum, YPM B03F0106.

---

[46] Edward L. Berthoud (1828-1910) was a founder, promoter, and builder of Golden, arriving in 1860 a year after the founding of Golden. He was a degreed civil engineer with a passion for railroads and natural science. Berthoud was the first to write about the geology of Golden in 1861 and collaborated on the geology of Golden with F.V. Hayden and Archibald Marvine from 1869 to 1874. Berthoud was the first instructor of Geology at CSM and a friend of, and probable mentor to, Arthur Lakes. In commemoration of Berthoud's passion for geology and his instrumental role in establishing the Colorado School of Mines in 1874, the Geology building at CSM was named in his memory in 1939.
[47] Sprain et al., 2018. Variation in precision is 8000 to 43000 years, which is extremely precise geologically.
[48] Dr. Walter Alvarez first advocated the impact theory in the mid-1970s.
[49] Yale Peabody Museum correspondence B03F0106 Berthoud to Marsh 1861 to 1883, Letter 4, June 21, 1874. The actual tooth is specimen YPM 4192.
[50] Carpenter and Young, 2002.
[51] Roland W. Brown (1893-1961) joined the USGS in 1929 as a paleo-botanist. He was charged with finding and resolving the K-Pg boundary location in the western U.S., a controversy that had existed since the 1880s. South Table Mountain was one of his field areas, and in 1943, he published the paper documenting the boundary location (Brown, 1943).

South Table Mountain, near the Denver West office complex, and only finding mammals and Paleogene plant pollen above it. Fast-forwarding to the 1990s, as the asteroid-impact theory became accepted, more research was conducted at the same site, trying to better locate the boundary.[52] Unfortunately, researchers discovered that the actual boundary was in a zone occupied by river deposits, in which an eroding river channel removed the unique impact deposits like the iridium layer. Those same researchers, however, re-confirmed the upward change of dinosaur to mammal fossils and a big change in plant life above and below the stream deposits. The boundary area is commemorated by the Cretaceous and Tertiary trails in the South Table Mountain Open Space park.

Fig. 32. West side of North Table Mountain with prominent lava flows 1, 3 and 4. The K-Pg boundary is in the hillslope above Flow 1. View to east.

Table Mountain Lava Flows

Within the Denver Formation, four separate lava flows of potassium-rich basalt called "shoshonite"[53] are present on North and South Table Mountain. The oldest Flows 1 and 2 are only present on North Table Mountain. Flow 1 ends just north of the steep trail on the west side of North Table Mountain (Fig. 32). Flow 2 is very localized, only having been recognized on the north side of North Table Mountain directly underlying Flow 3.[54]

Flows 3 and 4 form the prominent cliffs surrounding and capping both mountain tops. The most extensive Flow 3 was at least ten miles long, extending to the lowermost north flank of Green Mountain. Youngest Flow 4 is more restricted, only reaching from North Table Mountain to parts of South Table Mountain. Flows 3 and 4 also form the prominent peak of Castle Rock on South Table Mountain.

Modern age-dates of the basalt flows on North Table Mountain (Table 1)[55] show that the K-Pg boundary (66.0 ± 0.043 Ma) lies above lava Flow 1 (Fig. 32) and below Flows 2, 3, and 4. The dates have a high degree of precision. However, the accuracy of the date for Flow 4 is a problem.[56] Flow 4 is obviously the youngest of the four flows, being deposited on top of Flow 3, yet has a numerical age closer to that of Flow 2 (Table 1). This is a problem with the physics of numerical age-dating: an issue of precision versus accuracy.

Table 1. Ages of lava flows on North Table Mountain (Milliken et. al., 2018)

Lava Flow	Age	Error
Flow 4	65.9 Ma	±0.3 Ma
Flow 3	65.5 Ma	±0.3 Ma
Flow 2	65.8 Ma	±0.2 Ma
Flow 1	66.5 Ma	±0.3 Ma

As suspected by geologists since the late 1800s, a volcanic vent complex at Ralston Dike, about two miles north of North Table Mountain, is the likely source of the Table Mountain volcanic flows (inset on Fig. 21). The outer part of Ralston Dike is 65.4 ± 0.2 Ma old,[57] an age compatible with Flows 2 through 4 on North Table Mountain. Also, the basalts at Ralston Dike are the same composition as the Table

---

[52] Kauffmann et al., 1990.
[53] Iddings (1895, p. 943) named this rock where he first collected it at the headwater of the Shoshone River at Two Oceans Pass in the Absaroka Mountains of northwest Wyoming. The river is named after the Shoshone Indian tribe.
[54] Drewes, 2008.
[55] Milliken et al., 2018. These ages are very precise in geologic terms but some math wizard might conclude that they overlap. The order of the flows 1,2,3,4 (oldest to youngest) is correct but the exact timing with their ± variance will always have measurement error.
[56] A measurement can be precisely inaccurate. Flow 4 is the youngest flow, as it overlies the other flows. Thus, the older age date for Flow 4 is inaccurate, even though it a precise number, with a small measurement error. These issues drive a geologist crazy.
[57] Milliken et al. 2018.

Mountain basalts. The flows apparently came up along the subsurface trace of the Golden Fault during the Laramide Orogeny (Chapter 5).

Ralston Dike is more correctly called an exhumed volcanic plug. That is, it did not form a large volcano or an underground dike per se. The small cone where the vent came to surface was stripped away by erosion long ago, leaving only the eroded plug as the volcanic remnant. The lava flows from Ralston Dike flowed across the ancient landscape in a southerly direction toward present-day North and South Table Mountains, filling shallow channels of pre-existing rivers and streams.[58]

Hard-Luck Basalt

At Interplaza (Home Depot/Kohls) shopping center in south Golden (Fig. 21), construction excavations encountered the flat-lying basalts of Flow 3. As the story goes, the builders had **not** expected to encounter the volcanic rock because the USGS geologic map (Scott, 1972) did not show them, even though earlier workers had mapped them.[59] In the course of excavating, the builders struck hard rock before getting to the final grade needed for construction, requiring blasting. Unpleasantly surprised, the developers needed more funds to defray the excess construction costs. The hard-luck lesson learned here is that doing one's geology homework saves money in the long run.

Zeolite Minerals

Beautiful zeolite minerals on North Table Mountain have long been famous to mineral collectors.[60] Zeolites partially fill natural holes, called vugs, in the lava flows (Fig. 33). Zeolite minerals form in volcanic rocks by reaction with alkaline groundwaters, over short to long geologic time periods. Minerals include analcime, thompsonite, stilbite, and chabazite, some of which are on display at the Mines Museum of Earth Science. The minerals were first formally documented on North Table Mountain in 1882,[61] after Arthur Lakes and his students opened a small quarry in 1874 on the south side of the mountain, later the site of the Tramway Quarry.[62] Later, another small quarry opened on the east side of North Table Mountain, dug by faculty and students at CSM.[63] Currently, mineral collecting at the former Tramway Quarry is by permit from the Jeffco Open Space office.

Fig. 33. White zeolite crystals of analcime lining a 2.5-inch wide vug in brown basalt from North Table Mountain. Image by Dennis Gertenbach, used with permission.

---

[58] Drewes, 2008
[59] Noted by Marvine, 1874; reported by Cross in 1896; shown by Reichert, 1954; and re-reported by Drewes, 2008.
[60] See also "North Table Mountain" in Chapter 11.
[61] Cross and Hillebrand, 1882; also cited in Cross and Hillebrand, 1885, p. 452.
[62] Van Horn, 1976, p. 109. See Chapter 11 for Tramway Quarry history.
[63] Johnson and Waldschmidt, 1930, p. 118.

Castle Rock Geology

Castle Rock dominates the eastern view in Golden and serves as the main part of the logo for the City of Golden. Basalt Flows 3 and 4 form the cliffs of Castle Rock (Fig. 34). The prominent vertical lines in the cliff face are columnar joints formed when the lava cooled at the Earth's surface. Columnar joints are common features of basalt lava-flows around the world. Flow 3 is thought to have flowed down a pre-existing stream or river channel, creating the irregular contact at its base.[64] The contact between Flows 3 and 4 is a reddish-brown zone full of vugs.

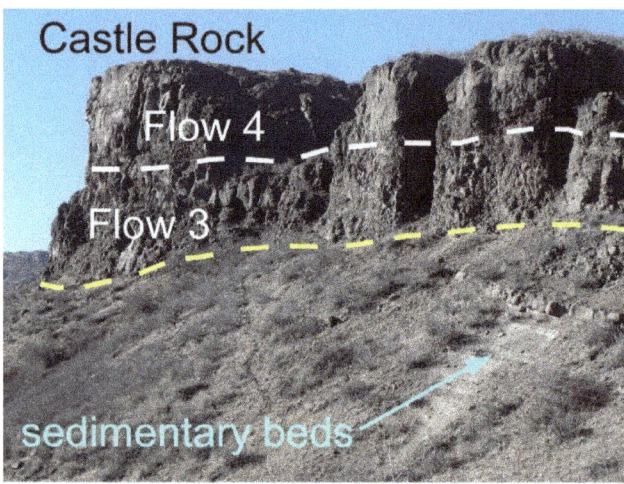

Fig. 34. Castle Rock consists of basalt Flows 3 and 4, overlying sedimentary beds of the Denver Formation. Dashed lines are contacts between flows (white) and the underlying sediments (yellow). View to northwest from Lubahn Trail.

Arthur Lakes and his students found fossil leaves and bones in the sedimentary beds below the lava flows south of Castle Rock. Many of those fossils are now in the collections of the Mines Museum of Earth Science, the Yale Peabody Museum in Connecticut, and the U.S. National Museum (Smithsonian) in Washington D.C. The leaves show that the land prior to the lava flows was forested with subtropical trees like palms, magnolias, and sycamores.

The K-Pg boundary location is 230-260 feet below the base of Castle Rock, just above the Lubahn trailhead on Belvedere Street (Fig. 21).[65] Therefore, standing at the trailhead, gazing up at Castle Rock, you are looking at about ½ Ma (from about 66 to 65.5 Ma) of geologic time. You can also consider that Earth-shattering day in Golden (and the rest of the world) when the asteroid created the K-Pg boundary just above your feet. Walking uphill toward the lava flows, recall that above the K-Pg boundary, mammals, not dinosaurs, walked through the subtropical environment of that time.

A prominent light-colored vertical scar on the southwest side of Castle Rock (Fig. 35) is the result of a rock fall that happened at 5:20 a.m. on 23 March 1958. The rock fall broke along the lines of the columnar joints in the cliff. The loud boom of the rock fall startled a passing newspaper delivery boy.[66] As the nearly 100-foot high rock column fell to the slopes below, it broke into smaller pieces that bounced and slid down the slope, creating divots and furrows in the dirt. One nearly 4-foot

Fig. 35. Rock-fall scar (arrow) from 23 March 1958 on Castle Rock. View to southeast.

---

[64] Drewes, 2008.
[65] The K-Pg boundary below Castle Rock is not precisely located, because basalt Flow #1 is not present on South Table Mountain.
[66] Newspapers, printed on paper, used to be home-delivered by pre-teenagers riding bicycles, by about 6 a.m. every morning!

long rectangular block from the fall moved 900 feet before stopping in the driveway of a nearby house at 400 18th Street.[67]

The Great Rock'n'Roll Caper

In this book we talk about mountain ranges rising and then completely eroding away over millions of years. But, on a human time scale, how do mountains actually break down? The wearing-down process is accomplished by streams eroding their banks and rocks falling off cliffs, among other things. But those events, like the rock fall from Castle Rock in 1958 (above) are infrequent on a human time-scale, and they are quickly forgotten.

One particularly ironic event happened in 1980 when a developer challenged a Jefferson County "no-build" policy on the north side of North Table Mountain directly below the basalt cliffs of Flows 3 and 4. Acknowledging the many large boulders lying around on the slopes, the developer believed that the boulders would only "creep slowly" down the slope rather than rolling, thus posing no threat to structures below the cliffs. An experiment was conducted to "prove" the "erroneous" notion that boulders could indeed roll downhill. People pushed several boulders down the hill from near the cliff base. The experiment quickly ended, however, when one boulder not only rolled far downhill but then bounced ten feet into the air, striking a cross-bar off of a high-tension power line at the base of the slope.[68]

How long does it take to wear down a mountain range, a few boulders at a time? A very long time in human time.

Fans of Green Mountain

Green Mountain dominates the skyline south of Golden. Although most of Green Mountain is City of Lakewood open space, the southernmost neighborhood of Golden, Golden Heights, occupies the north flank of the mountain (Fig. 21). Walking and mountain-biking trails abound on Green Mountain within the beautiful 2400-acre William Frederick Hayden Park (not the same Hayden as the 1800s geologist). Views from the mountain top are spectacular.

Rocks of the Green Mountain Formation are about 65 to 64 Ma[69] old, which is about 500,000 years younger than Table Mountain lava Flow 3. They are mostly conglomerate interbedded with lesser amounts of sandstone and siltstone. Cobbles and boulders in the conglomerate are a variety of rock types suggesting erosion from the still-rising Front Range uplift.[70] Also, Green Mountain Formation pebbles and boulders (Fig. 36) are larger than most of those in the older Arapahoe and Denver formations. Larger sizes point to much faster and more voluminous river flows than those depositing the Arapahoe and Denver formations. Geologists think that the Green Mountain Formation represents a series of

Fig. 36. Mixed conglomerate sizes in the Green Mountain Formation near the summit suggest deposition across an alluvial fan. The small blue pencil to right of large boulder is about 5 inches long. View to

large alluvial fans that spread out of the nearby Laramide moutains immediately to the west.[71] The interbedded sandstone and siltstone of the Green Mountain Formation have yielded many plant and mammal fossils, but no dinosaurs, as the rocks formed almost two million years after the K-Pg extinction.

---

[67] Van Horn, 1976, p. 89-90.
[68] Noe et al., 1999, p. 13.
[69] Obradovich, 2002, p. 168-169, dated a volcanic ash bed on the summit of Green Mountain at 63.94 ± 0.28 Ma.
[70] Raynolds and Johnson, 2003.
[71] Dechesne et al., 2011, Plate 13.

Chapter 5: Mountains UP!

Imagine a 1000+ foot-high mountain range rising up along Illinois Street in the middle of Golden. That is what Golden might have looked like 67-66 Ma ago during what is called the Laramide Orogeny. All the steeply tilted beds west of Illinois Street were part of a Laramide mountain range. Everything east of Illinois Street was "flattish," or gently tilted, like the rocks on North and South Table Mountains. Hence the Laramide Mountain Front was right along Illinois Street (Fig. 37), instead of where the Front Range is located today. What happened? Around 67 Ma ago, Golden got squashed by deep, laterally-

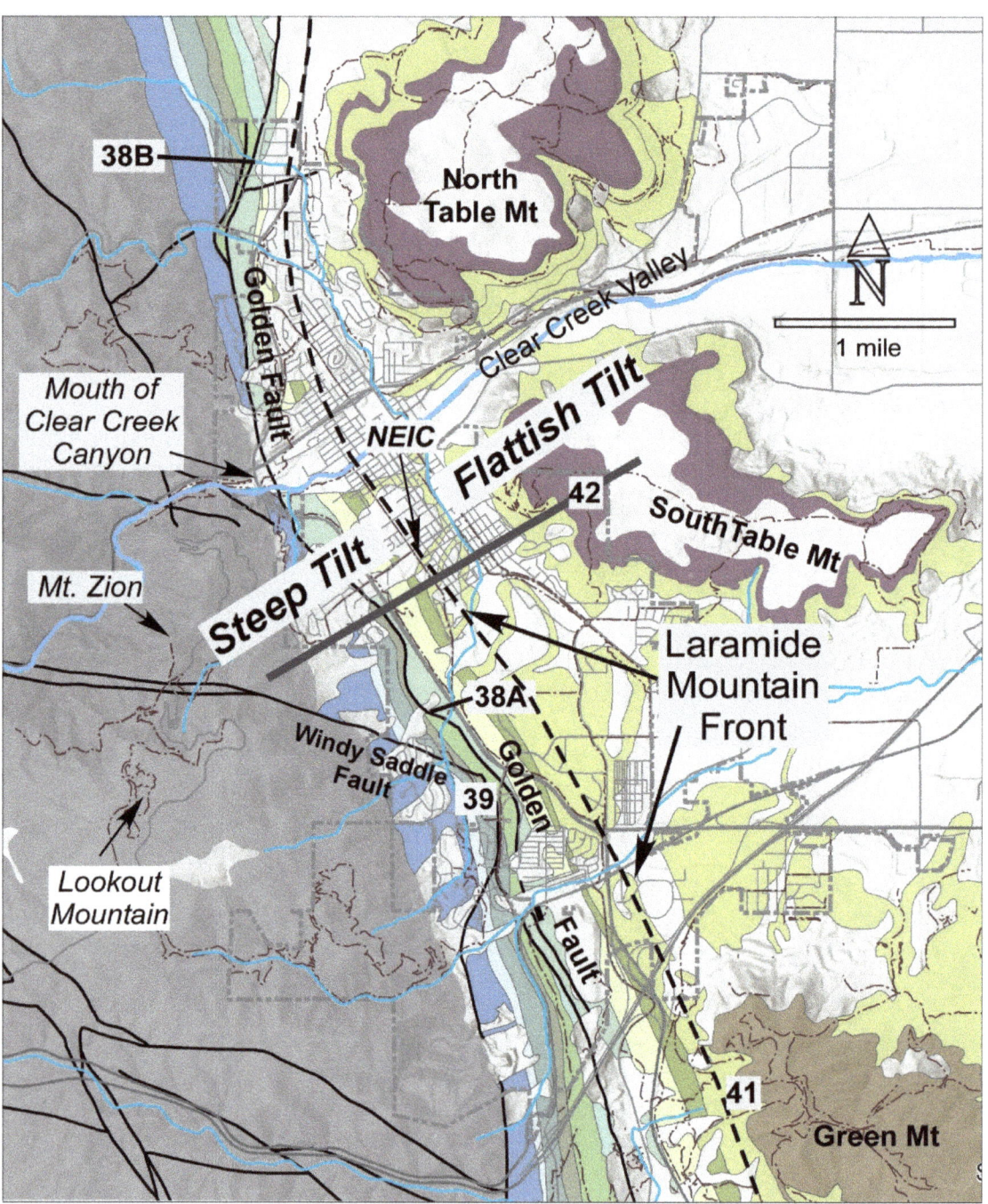

Fig. 37. Structural geologic features of the Golden area. Dashed black line is the Laramide Mountain Front. Numbers correspond to Figure locations. Geologic units as in Fig. 4. NEIC is the location of the USGS National Earthquake Information Center on Illinois Street.

directed crustal forces. All the Precambrian bedrock and sedimentary rock formations on the central and west side of Golden got folded, faulted, and tilted up as the mountains came up. At the time, earthquakes likely happened right in town. Exciting to a geologist, for sure, but maybe not so much to a home or business owner. That is the short story. The longer story of how and why Golden got squashed in the Laramide Orogeny unfolds below.[72]

Squashing Golden

The major character in the drama of squashing Golden is the Golden Fault System. It consists of a structural deformation zone including all the steeply tilted beds on the west side of Golden, the Golden Fault, the Windy Saddle Fault (Fig. 37), and many faults too small to show on a map. Historically, the Golden Fault System is thought responsible for uplifting the entire Rocky Mountain Front Range high above the pre-existing land surface of the time. The "flattish," slightly tilted beds of the two Table Mountains and Green Mountain hold evidence of later and more subtle squashing as the Laramide Orogeny waxed and waned from 66 Ma onward.

Severe Squashing

The tectonic forces responsible for squashing Golden were compressive forces transmitted laterally, or sideways, across Western North America. You can model it in following way. Take a small towel, put it on the floor, and push from both ends. The two-way push on the towel-ends is compression. In the middle of the towel, folds will form. If you keep pushing, the folds will flop over, as in overturned folds. A super-stiff, or brittle, towel might break, like a fault. If you kept pushing across the break, one side would ride up over the other, which is like a reverse fault. Severe geologic squashing forms a lot of structural complexity, including reverse faults and overturned folds. And, the west side of Golden was SEVERELY squashed. All of the steeply tilted beds on the west side of Golden (Fig. 37) are part of the squashing episode that includes reverse faults and overturned beds.

Golden Fault

Most geologists consider the Golden Fault to be the major fault in the Golden Fault System. On the ground surface, the Golden Fault has a winding pattern (Fig. 37). It becomes curved to the west through central Golden, where it is closest to today's Front Range at the mouth of Clear Creek Canyon. The Golden Fault places Early Cretaceous and older rocks on the west, on top of much younger, uppermost Pierre Shale on the east. At the mouth of Clear Creek Canyon, it places the basal Pennsylvanian Fountain Formation next to the upper part of the Cretaceous Pierre Shale, marking the location of the biggest difference in rock-age across the fault. These patterns are characteristic of a reverse fault,[73] a geologic fault in which rocks above the fault plane move over top of the rocks below.

The earliest geologists observed that the Dakota Hogback "disappeared" through central Golden but did not understand why. They further did not understand why most of the Benton Group and of the Pierre Shale (see Geologic Column, Fig. 4) did not exist in Golden, but those rocks were exposed north and south of town. By 1900, most geologists thought that a fault was responsible for this disappearance, but at least one author called it a "mysterious fault."[74] In 1917,[75] a CSM Professor of Geology had the insight that a reverse fault, named the Golden Fault, was the reason why the rocks of the Dakota Hogback, the Benton Group, and 8000 feet of Pierre Shale had "disappeared" in western Golden. Rocks on the west side of the Golden Fault broke off and rode up and over those younger Cretaceous rocks, like your super-stiff "towel" model. At the same time, the Laramide Orogeny thrust up the Front Range mountains.

---

[72] If you are not a geologist, you should review "Geologic Structures" in the Geology Primer before reading the "long story."
[73] It is also a thrust fault, as we explore below.
[74] Geijsbeek, 1901.
[75] Zeigler, 1917.

*Fig. 38. The Golden Fault (dashed line) abruptly cuts off rocks of the Dakota Hogback and puts the Dakota **up** and over the much younger Pierre. All the Benton Group, Niobrara Formation, and most of the Pierre Shale are missing at the surface because of the fault. A. Eagle Ridge in south Golden (view to south). B. Pine Ridge in north Golden (view to north).*

At Kinney Run Gulch (Fig. 38A), the Golden Fault cuts off the Dakota Hogback (Eagle Ridge). In north Golden, the same pattern exists where the Dakota Hogback (Pine Ridge) emerges abruptly at the Golden Fault (Fig. 38B). In both places, the fault curves around, first placing the Benton and Dakota rocks over the Upper Pierre Shale, then cutting off the Benton and Dakota rocks and placing the Jurassic Morrison Formation on top of the Upper Pierre Shale. Older rocks are consistently placed next to younger rocks along the Golden Fault. Studies have shown that in places, the Golden Fault plane dips westward at varying, but relatively low, angles.[76] A reverse fault that dips at a low angle is called a thrust fault, which can be applied to the Golden Fault. The same studies have shown, as do the outcrops of all the rock formations along the Golden Fault, that the rocks on both sides of the Golden Fault are vertically tilted to overturned: a whole lot of faultin' going on!

Windy Saddle Fault

The Windy Saddle Fault is in southwest Golden, west of the Golden Fault (Fig. 37). Starting near Apex Gulch, the trace of the Windy Saddle Fault curves eastward, turns north, and then curves northwestward at the historic Cambria Kiln (location "Y" on Fig. 39), where it is well-exposed across

Fig. 39. The Windy Saddle Fault cuts off the Glennon Limestone Member of Lykins Formation at "X." It then bends the white Lyons Formation outcrop to the northwest at the Cambria Kiln at location "Y." June 1931 aerial photograph, Lowry Air Force Base, Roll 25 Frame 32, courtesy of Arthur Lakes Library CSM.

---

[76] Dames and Moore, 1981, trench logs.

Kinney Run Trail. At the Cambria Kiln, the Windy Saddle Fault cuts off the north-trending Glennon and Falcon limestone members of the lower Lykins Formation (location "X" on Fig. 39) and bends, or folds, the rocks of the Lyons Formation to the northwest. On the west side of the CSM survey field, the fault trace enters the Precambrian basement rocks of the Front Range and crosses Windy Saddle (its namesake) at the trailhead between Mount Zion and Lookout Mountain. Is the Windy Saddle Fault also a reverse fault? Yes, it always places slightly older rocks over younger rocks, but the offset is much less than that of the Golden Fault, making the Windy Saddle Fault a less important fault, but still part of the Golden Fault <u>System</u>.

The Windy Saddle Fault is also a good example of how a fault influenced historic mining. Limestone mining[77] trenches within the Glennon limestone in south Golden (Fig. 39) followed the outcrop along today's streets of Berthoud Way and Somerset Drive. Because the Windy Saddle Fault cuts off the Lykins limestone at the Cambria Kiln, the miners had to move about ½ mile to the northwest, over to today's CSM survey field, to find the continuation of the limestone beds.

Other Faults

Myriad small-offset reverse faults are visible in many of the outcrops of the Laramie Formation and Dakota Group in west Golden, some of which have slickensided surfaces, especially in sandstones of the Laramie Formation, like those visible along the Weimer Geology Trail at CSM (Chapter 4, Fig. 27). Not only did early miners find minor faults like the Windy Saddle Fault, but they found them in the clay mines along the Dakota Hogback; in the Laramie Formation; and in the coal mines near Clear Creek, among other places. Such faults were the curse of miners, because miners had to spend extra effort and money to dig tunnels trying to find the faulted-out continuations of the coal or clay seams. Oddly, though, they found that some of those minor reverse-fault-planes tilted, or dipped, eastward, while others dipped westward, making it difficult to predict where to find the offset coal/clay seams.

One such east-dipping fault plane ended the operation of the Santa Fe Mine (Chapter 10) along Eagle Ridge. Around 1900, G.W. Parfet dug a new adit, or tunnel, into the west side of Eagle Ridge that completely missed the Dakota fire-clay seam that he sought to find.[78] The clay seam had been moved, or offset, by an east-dipping reverse fault: totally unexpected! Many decades later, geologists struggled with why so many east-dipping reverse faults existed along the Golden Fault System.

Gentler Squashing

Rocks <u>east</u> of the Laramide Mountain Front, including North and South Table and Green Mountains, are "flattish" (Figs. 37 and 40), and hence, are not nearly as squashed as those on the west side of Golden. Why? Because the relatively stable expanse to the east formed a "backstop," like one edge of your "towel" model for structural deformation. Golden was and still is on the eastern side of Western North America. Severe deformation across all of the western U.S. reached the end of the line so-to-speak, right in the middle of Golden.

Fig. 40. "Flattish" North and South Table Mountains are only tilted 3 to 5° down to the southeast (dashed line). The mountains were originally continuous. View to northeast.

---

[77] Chapter 11 discusses limestone mining in the Glennon Member and the history of the Cambria Kiln.
[78] Patton, 1904.

Still, a little structural deformation happened east of the Laramide Mountain Front, affecting the flattish, or slightly tilted rock layers of North and South Table (Fig. 40) and Green Mountains. The rocks deposited on North and South Table Mountains show evidence of subtle tilting as part of the continuing Laramide Orogeny during and after deposition. For example, although originally deposited across a nearly flat landscape, forces acting after deposition tilted all of the rocks on North and South Table Mountains down to the southeast at 3-5° (Fig. 40). That tilt is enough that the most extensive of the lava flows, Flow 3, ends up cropping out at the base of Green Mountain at the Jeffco Fairgrounds and the Interplaza/Home Depot shopping center (Chapter 4).

On Green Mountain, the rocks above lava Flow 3 add another 1000 feet of combined Denver and Green Mountain formations to the geologic story. Those rocks are tilted at varying angles on Green Mountain. The Denver Formation is tilted around 30° (Fig. 41), which is a big change from vertical in the underlying Arapahoe, Laramie, and Fox Hills rocks. The Green Mountain Formation is only tilted 10.° The different tilt angles represent separate times of tectonic tilting during deposition of first the Denver and then the Green Mountain formations from about 66 to 64 Ma.

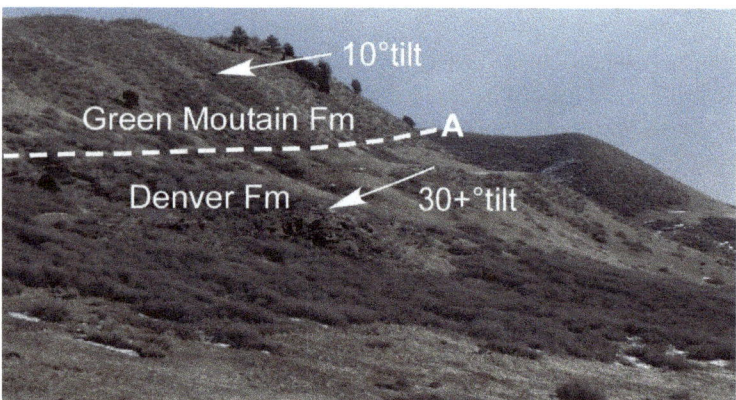

Fig. 41. A 20-degree tilt difference between the Denver and Green Mountain formations (dashed line A is contact between them) on the western slope of Green Mountain forms an angular unconformity (after Drewes and Townrow, 2005). View to southeast from Green Mountain Trail in William Hayden Open Space.

Below the Ground

For geologists, the ultimate challenge is predicting what is below the ground, be it 5 feet, 20,000 feet, or to the center of the Earth.[79] And, what the Golden Fault System looks like below the ground is a thought-provoking example of this challenge. As mentioned above, even recognizing the presence of the Golden Fault System took decades, beginning when 25-year-old Archibald Marvine rode horseback into Golden in 1873 with the assignment to make the first geologic map of a 5000 square-mile area encompassing much of northeastern Colorado. Marvine mapped the distribution of rocks on the ground surface in Golden and speculated that the Dakota Hogbacks, for example, might be interrupted by a fault,[80] at a time in geologic understanding when the concept of thrust faults was controversial. Over 30 years later, CSM Professor Victor Zeigler recognized the Golden Fault as a west-dipping reverse fault.[81] Early miners recognized many of the minor faults in Golden, including the Windy Saddle Fault. The sharp change in tilt across Illinois Street (Fig. 37) was recognized early-on also, but not named until the mid-1990s, when it was called the "Basin Margin Fault."[82] We call this feature the "Laramide Mountain Front" (Figs. 37 and 42). Until recently, geologists ignored the oddly east-dipping reverse faults, such as those on Eagle Ridge and in the Laramie Formation along the Weimer Geology Trail. Geologist Ned Sterne broadly looked at the data along the Front Range from Wyoming to Colorado Springs and

---

[79] Why? For example, building foundations (like the basement of your house), earthquakes, groundwater, mineral resources, and hydrocarbons are all "below-the-ground." Soil and rocks change dramatically in complicated ways, so predicting them is essential and occupies a lot of what geoscientists do.

[80] Marvine, 1874, p. 136-137. He thought it would take "a few hours work" to clear up the problem. It is still debated!

[81] Ziegler, 1917, finally published an article saying it was a fault. Marvine (1874) and Emmons et al. (1896), called it an unconformity, known as the "Golden Arch," not to be confused with the modern fast-food restaurant. Zeigler's insight was greatly appreciated by the geologic community in and around Denver.

[82] Weimer, 1996, and Weimer and Ray, 1997.

proposed a different and more complex interpretation of the Golden Fault System, using a concept of triangle zones coupled with back-thrusts.[83] Such features are common at the edges of "severe squashing" zones, such as exist along most mountain fronts around the world.

Regardless of how one interprets fault complexities, a basic truth remains. If you stand with your hand on the contact between the Fountain Formation and the Precambrian basement rocks, you would have to drill a hole around 12,000 feet deep to again encounter the Precambrian basement below-the-ground (Fig. 42) beneath your feet. The bottom line is that movement along the Golden Fault System shoved the mountains up and over the sedimentary basin to the east.

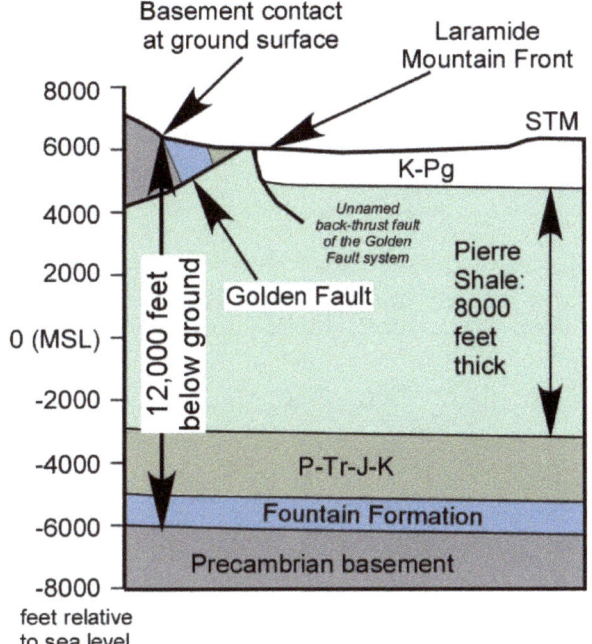

Fig. 42. Generalized structural cross section across Golden, location on Fig. 37. Fountain Formation (blue), P-Tr-J-K (dark green) includes Lyons through Niobrara formations, Pierre Shale (light green), and K-Pg (white) includes Fox Hills through Basalt Flow 4 of the Denver Formation. STM is South Table Mountain. Vertical and horizontal scales are the same. Authors' interpretation after Sterne (2006, 2017).

After 64 Ma in Golden

What happened after 64 Ma in the Golden area? The Laramide Orogeny lasted a lot longer, ending around 45 Ma,[84] but the rock story in Golden is incomplete. We know that North Table Mountain was the site of a volcanic intrusion at 46.94 ±0.15 Ma (Late Eocene).[85] Other than that, though, we need to look at other rocks to tell the rest of the story, including those rocks exposed around the town of Castle Rock in Douglas County, 32 miles southeast of Golden.[86]

Younger Paleogene rocks in Douglas County tell a story of deposition by energetic rivers issuing from the still-rising Laramide Mountain Front. Alluvial fans, younger than those of the Green Mountain Formation, kept forming next to the Laramide Mountain Front. Periodically, thick, clay-rich soil horizons accumulated across the Denver area during times of little or no deposition, reflecting "resting" periods across the landscape. Such periods of relative land-surface stability formed a broad plain that extended across the Laramide Front Range and the adjacent Denver Basin, similar to the Great Plains of today.[87]

A final period of volcanism and river deposition from 37 to 34 Ma further buried the landscape. Around 37 Ma[88] a violent volcanic eruption west of the Front Range covered part of the landscape in the greater Denver area with tens of feet of volcanic ash flows, killing everything that lived above ground. After that eruption, rivers seemingly went "wild," depositing thick sediments in deep valleys. Indirect evidence suggests that at least one of those rivers deposited sediments across the Golden area.[89]

---

[83] Sterne, 2006, 2017. Discussions regarding the Golden Fault System with Ned Sterne and CSM Geology Professor Bruce Trudgill are highly appreciated.
[84] Some geologists place the end of the Laramide Orogeny at 55 Ma and have good reasons to do so. Here, we extend the time to 45 Ma to include, for example, the late volcanic phase on North Table Mountain.
[85] Milliken et al., 2018: the young age initially caused controversy within the geologic community, but the age-date is accepted.
[86] Dechesne et al., 2011, and Raynolds and Johnson, 2002.
[87] Cather et al., 2012.
[88] The Wall Mountain Tuff 36.7 ± 0.07 Ma (Chapin et al., 2014) erupted from a caldera near Mount Princeton in Park County, CO.
[89] Koch, et al. (2018) show good evidence for a river crossing Golden, but those deposits were later eroded away.

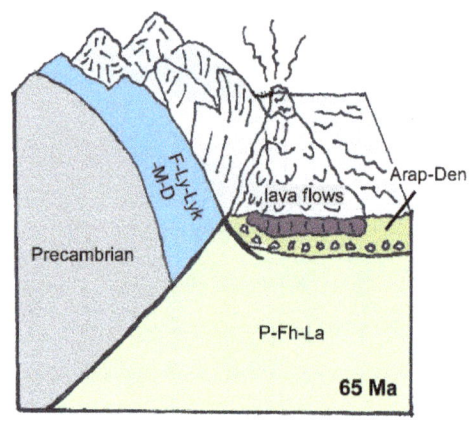

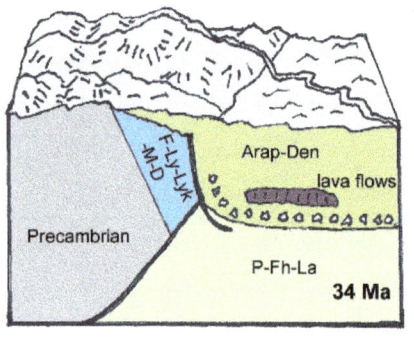

Rocks equivalent to the Green Mountain and Denver formations are known in the subsurface (underground) of the Denver Basin to the east, as per data from wells drilled for groundwater, and oil and gas. Studies tell us that as much as 3000 feet[90] of sedimentary rock, in addition to the 1000 feet between Basalt Flow 3 and the top of Green Mountain, was deposited across Golden and Denver between about 65 to about 34 Ma (Fig. 43). But by 34 Ma, the Laramide Rocky Mountains west of Golden became worn down to a series of low hills that were partly buried by sediment (Fig. 43). At that point in time, about 4000 feet of sedimentary rocks used to be on top of North and South Table Mountains! That was quite a change from 66-65 Ma when the high Laramide Mountain Front was along Illinois Street, and the Table Mountain basalt-lavas flowed across the east side of town.

Fig. 43. Schematic (not to scale) diagrams show the landscape around Golden at two times. First, is at the peak of the Laramide Orogeny in Golden (about 65 Ma) when the mountain front was at Illinois Street, and lava flowed down the east side of the mountain front. Then, by about 34 Ma Golden became buried, placing the Table Mountain basalt under about 4000 feet of sediment. Colors represent different formation. Blue=Fountain, Lykins, Lyons, Morrison-Dakota. Bright green=Pierre-Fox Hills-Laramie. Yellow-green=Arapahoe-Denver. Brown=Table Mountain lava flows.

Ultimate Cause of Squashing

Broadly speaking, what caused the Laramide Orogeny? The answer lies in understanding the plate tectonics of North America. In Late Jurassic time, a massive oceanic plate began subducting under the western side of the North American continent, being pulled under today's states of Calfornia and Nevada. Originally tilted steeply down into the Earth's mantle, the subducting oceanic plate began to flatten until it became joined to the base of the Earth's crust in the Late Cretaceous through Paleogene. The flattening and joining process transmitted huge compressive tectonic forces through the Earth's crust that squashed the entire Western U.S., from California to eastern Colorado. The result was an extensive series of up-faulted mountain ranges and adjacent down-dropped basins. Collectively called the Rocky Mountains, reaching from New Mexico to Canada, the Front Range of Colorado was one of those mountain ranges. The entire mountain-building episode was called the Laramide Orogeny. Some of the best evidence for that orogeny happens to be in Golden, Colorado, as told by the stories of the Fox Hills, Laramie, Arapahoe, Denver, and Green Mountain formations (Chapter 4), and the Golden Fault System.

Modern Earthquakes in Golden?

Earthquakes may have been common in Golden during the Laramide Orogeny, but in the recent geologic past, Golden has been seismically uneventful. While no earthquake epicenters have been recorded in Golden,[91] earthquakes have been felt here. Some long-time Golden residents may recall earthquake damage at the Golden Ridge condominiums east of Heritage Road in the early 1980s. On 18 October 1984 a magnitude 5.4 earthquake, centered in the Laramie Mountains of Wyoming, over 180 miles north of Golden, caused foundation failure, numerous wall cracks, and a gas leak.[92] Damage in

---

[90] Higley and Cox, 2007, p. 14.
[91] USGS earthquake catalog: https://earthquake.usgs.gov/earthquakes/search/
[92] Langer et al., 1991.

Golden was highly unusual for an earthquake that far away. The earthquake waves, coming from a focus 13 miles below the ground surface, seemed to be aimed right at the condominiums.

The largest historic earthquake known to have occurred in Colorado was on 7 November 1882. It was felt over a large part of Wyoming and Colorado. As reported for Golden, excitement already generated by a contentious election that day was further elevated by a "violent shock, in which the babble of voices simultaneously ceased. Strong men reeled and tottered…reaching out as if to grasp some support." [93] Structural damage was not reported in town. Long before the days of seismographs, the estimated magnitude was about 6.5, with the epicenter in the mountains west of Fort Collins, based on shaking and damage reports.[94]

Could the Golden Fault cause an earthquake now? Probably not. In the late 1970s, questions about the activity of the Golden Fault arose due to seismic-safety concerns at the then-operating Rocky Flats Nuclear Plant situated between Boulder and Golden. To evaluate the concerns, a major engineering firm, Dames and Moore, made a comprehensive study of the Golden Fault in 1980, under contract to the U.S. Department of Energy. The engineering firm dug several backhoe trenches across the Golden Fault, one of which was a few hundred feet north of Kinney Run Gulch in south Golden (below the bottom of the photograph, Fig. 38A). That trench showed that the Quaternary deposits of the Verdos Alluvium, more than 410,000 years old (Chapter 6), buried the bedrock fault-trace with no offset in the younger alluvium.[95] Since the Golden Fault has not been active in the last 410,000-plus years, it is considered a pretty dead fault.

National Earthquake Information Center

Golden is home to the National Earthquake Information Center (NEIC) at 1711 Illinois Street on the CSM campus (Fig. 37). The NEIC is part of the U.S. Geological Survey and was established in 1966.[96] The "national" in its name is a misnomer of sorts inasmuch as the NEIC tracks earthquakes around the world 24/7. It is also home to Golden's own seismograph station since 1974.

The NEIC has three main missions. First, it determines the location and size of all significant earthquakes and quickly disseminates this information nationally and internationally. Second, it collects and maintains an extensive seismic database as the foundation for scientific research, made possible by global seismograph networks. Third, the NEIC studies earthquakes to improve the ability to locate and understand them. Much of the earthquake information that you get in news, on your phone, and through social media comes from data servers right in Golden. The NEIC also conducts free public tours, but you need a reservation: call +1-303-273-8500.

---

[93] CT 8 November 1882.
[94] Colorado Division of Homeland Security and Emergency Management, 2021.
[95] https://earthquake.usgs.gov/cfusion/qfault/show_report_AB_archive.cfm?fault_id=2324&section_id= The USGS Quaternary Fault Database and Dames and Moore, 1981.
[96] https://www.usgs.gov/natural-hazards/earthquake-hazards/national-earthquake-information-center-neic The USGS NEIC website.

Chapter 6: Rockies Rebirth

You might say, "Wait a minute. I thought we just built the Rocky Mountains in the Laramide Orogeny." True enough, but geologists have come to appreciate another long page in the story. The Front Range we see today is a re-birth of the Laramide Rockies, a process that began about 34 Ma ago.[97] It is a story of erosion, finally carving out the modern landscape. It is also the reason why 4000 feet of rocks became stripped off all of northeastern Colorado, including 4000 feet of sedimentary rocks that used to be on top of North and South Table Mountains.

From about 34 to about 25 Ma, flat plains lay atop the buried Laramide mountain ranges throughout the Rockies. Rivers meandered across those plains, depositing sediments, initially with little downcutting. But, from 25 Ma up to the present, the huge region including Colorado and most of the western U.S. went through several broad uplift periods.[98] With each period, rivers became rejuvenated as gradients increased. Massive erosion stripped off thousands of feet of Paleogene sediments, essentially "un-burying," or exhuming, the earlier Laramide landscape. As any given river eroded down, it encountered a buried mountain range. Once trapped and carved into canyons within the former mountain ranges, the new rivers never changed course. They kept cutting down, crossing all rock layers and geologic structures encountered. Such rivers became known as "superimposed" rivers.[99]

Clear Creek Shapes the Golden Landscape

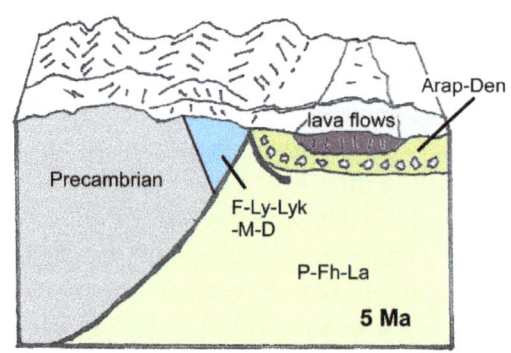

The Rockies Rebirth story in Golden is about Clear Creek and its tributaries. The first "ancestral" Clear Creek exhumed (un-buried) the Front Range bedrock and the basalt flows of the Table Mountains. After that, "modern" Clear Creek, flowing west to east, carved out Clear Creek Canyon and separated the two Table Mountains (Fig. 44). Two tributary streams, Tucker Gulch and Kinney Run Gulch, further eroded the Golden area into what became the north-south trending Golden Valley between the Front Range and Table Mountains. Step by step, Clear Creek and its tributaries shaped the modern landscape.

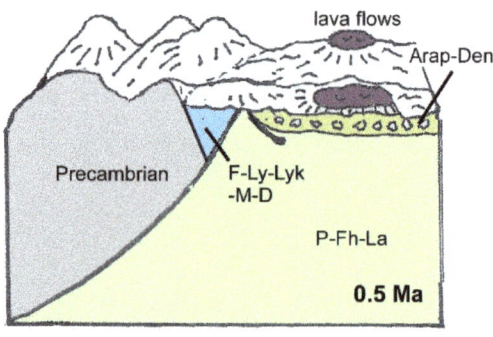

Fig. 44. Schematic (not to scale) diagrams show the landscape around Golden at 5 Ma when Ancestral Clear Creek flowed across the Table Mt basalt flows but had not yet incised through them. By about 0.5 Ma, modern Clear Creek had incised a deep canyon through the Front Range, separating the two Table Mountains. By 0.5 Ma, today's landscape in Golden would be recognizable. Colors and abbreviations as in Fig. 43.

---

[97] An engaging account of the re-birth of the Rockies is in Keith Meldahl's (2011) book in the Chapter titled "Range-Roving Rivers," p. 165-197. Meldahl's book is another "great read" about the geology of the Western U.S. with an emphasis on plate tectonics.
[98] Recounted by Cather et al., 2012.
[99] Archibald Marvine (1848-1876) was credited for this concept in John Wesley Powell's eulogy on Marvine's untimely death at age 25 (Powell, 1876). A good example of a superimposed stream is the Green River where it cuts straight across Split Mountain in Dinosaur National Monument, near Vernal, Utah. Another good example is the Dolores River where it cuts across the Paradox Valley in western Colorado near the town of Bedrock. The valley is so-named because of the "paradox" of the river cutting straight across the trends of geologic structures, instead of following those trends.

Ancestral Clear Creek

For the first part of the Clear Creek story, you must tap into your imagination. You need to "fill up" the Golden Valley to the elevation of the tops of the two Table Mountains, burying everything we see today. Standing atop North Table Mountain a few hundred yards from the west edge and looking toward today's mountains (Fig. 45) helps you imagine what the landscape may have looked like about 5 Ma ago, when the Golden Valley and Clear Creek Canyon did not exist. A low-relief upland existed across the Front Range. The Paleogene sedimentary cover had been eroded to expose the youngest Table Mountain lava flows, but the North and South Table Mountains were still connected. Likely, a stream system representing "ancestral" Clear Creek meandered across the landscape before eroding what we know today as the Golden Valley and Clear Creek Canyon (Fig. 44).

Fig. 45. View west from the top of North Table Mountain showing the rolling low-relief upland above 7500 ft. Range, a view perhaps little changed from about 5 Ma ago when the Golden Valley did not exist: it was filled with sedimentary rocks.

After 5 Ma,[100] Clear Creek cut down through the resistant Table Mountain lava flows, separating the two Table Mountains into north and south portions (Fig. 44). At this critical time, Clear Creek became a superimposed stream. Locked into its path between North and South Table Mountains, Clear Creek began downcutting its steep modern canyon. Because they were eroded at the same time, the inner gorge of Clear Creek Canyon and the Clear Creek Valley between the two Table Mountains have the same depth of about 600 feet. One could say that after 5 Ma, the modern path of Clear Creek was "set in stone."

Ice-Ages and Clear Creek

As Clear Creek began to cut its modern path, the entire State of Colorado began uplifting again, eventually reaching its current <u>average</u> elevation of 6800 feet, the highest of any State in the U.S. As before, uplift helped revitalize streams like Clear Creek by increasing erosive power. Eventually, the increased elevation of the Rocky Mountain Front Range and the cold climates of the Ice Ages caused glaciers to form. Along the 11,000-foot summits at the Continental Divide, glaciers carved out broad, deep valleys. Runoff of glacial meltwater, during times of de-glaciation at the ends of the several Ice Ages,[101] kept Clear Creek in a state of nearly continual downcutting.

Sloping Land-Surfaces and Alluvium in Golden

The modern landscape within Golden is the result of climate change during the Quaternary Period (2 Ma to present). Generally speaking,[102] wetter climate times during the Ice Ages provided increased weathering to form sediment and increased water volumes that moved sediment out of the mountains, depositing it downstream in the flatter, broader areas east of the mountains. Drier times,

---

[100] The timing is imprecise and debated among geologists. But the >1.5 Ma Rocky Flats Alluvium (Table 2) is 200 feet below the top of North Table Mountain, which means that North Table Mountain had already been eroded before the Rocky Flats Alluvium and surface formed.

[101] Retreat and advance of glaciers occurred at least four and up to 12 times in the headwaters of the Front Range over the last two million-plus years, but the only remaining evidence is for the last two times, during the Bull Lake and Pinedale glaciations. Globally, up to 30 glacial periods have been recorded since the Miocene Epoch (25-23 Ma) of the Neogene Period. See Geology Primer for the Neogene time scale, for which almost no rocks exist in Golden.

[102] This is a simplified concept. It is more complicated than that and is a topic of discussion among Quaternary geologists and geomorphologists.

during interglacial periods,[103] were times of relative landscape stability leading to forming sloping land-surfaces on top of thin, underlying sedimentary deposits. Quaternary geologists have mapped the youngest geologic units in Golden (Fig. 46), which are a combination of sloping land-surfaces and thin deposits.

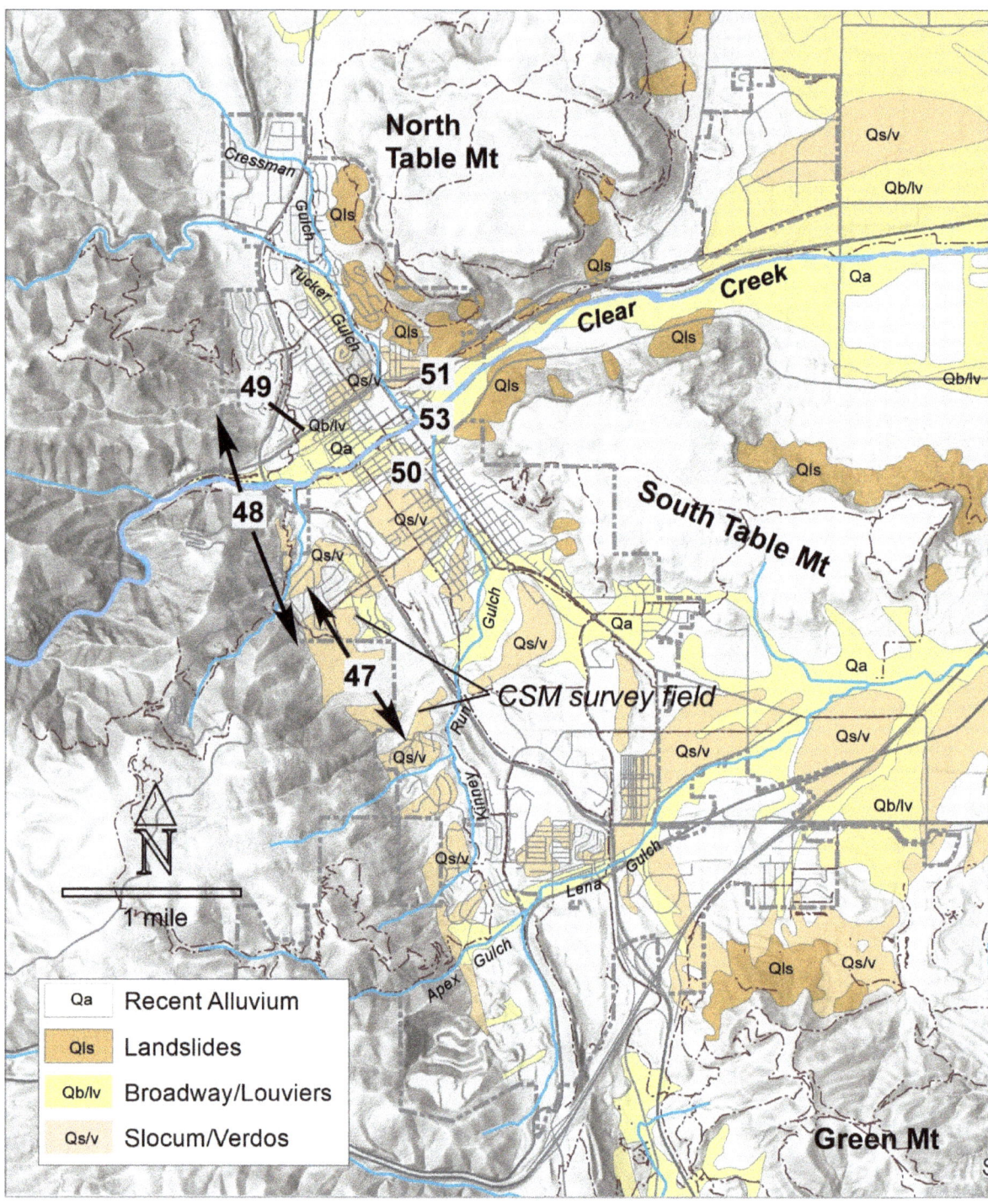

Fig. 46. Selected Quaternary geologic units in Golden, modified from Kellogg et al., 2008. See Table 2 for more explanation. Numbers correspond to Figure locations.

---

[103] We are currently in an Interglacial period.

In western and central Golden, two major sloping land-surfaces are present: the Verdos and Slocum surfaces, together ranging in age from 700,000 to 250,000 years old, respectively (Fig. 47, Table 2). They occupy higher elevations around 5900-6050 feet. They are best-preserved on the west side of the city, especially across the CSM survey field, west of US Hwy-6 (locality 47 on Fig. 46). Many of these

Table 2. Types of Quaternary alluvium in the Golden area, after Kellogg et al. (2008). Symbols: < means "less than;" > means "more than;" ≤ means "equal to or less than; ~ means "approximately;" ka is "thousand-years."

Alluvium Unit	Age Range (ka)	Elevation in Golden	Notes
"Recent"	< 13	in Clear Creek (CC) channel	Piney Ck & Post-Piney Ck of Kellogg et al. 2008
Broadway Alluvium	30-13	Low Banks above CC channel	Pinedale Glaciation
Louviers Alluvium	170-120	High banks above CC channel	Bull Lake Glaciation
Slocum Alluvium	400-250	≤5900 ft elevation	pre-Bull Lake Glaciation
Verdos Alluvium	675-410	≤6000 ft elevation	Lava Creek B ash (~631 ka) at base
Rocky Flats Alluvium	> 1,500	200 feet below North Table Mountain. Rocky Flats, north of Golden	formed after Clear Creek incision

Fig. 47. Sloping land-surfaces on the CSM survey field are covered by Verdos (V) Alluvium Lower slopes in foreground are covered with Slocum (S) Alluvium. The surfaces and thin alluvial deposits lie on top of bedrock. View to southwest. Lookout Mountain is highest peak on the skyline.

gently sloping surfaces are now covered with homes, as they have great views and are easy to develop.

Each sloping land-surface has a thin layer of sediment, called alluvium, beneath it. Commonly the sediment consists of a mixture of boulders, cobbles, sand and silt, all called "surficial deposits" because they exist right at the ground surface, covering the deeper bedrock. These unconsolidated deposits have never been buried deep enough to become rock.

The Verdos Alluvium contains an important volcanic ash, the "Lava Creek B" ash, that blew out of a Yellowstone super-volcano eruption about 631,000 years ago, spreading over the entire western and central U.S.[104] Occurring at the bottom of the Verdos Alluvium, the Lava Creek B ash tells us that the overlying Verdos surface is younger than 631,000 years. Quaternary geologists estimate that the Verdos surface is about 410,000 years old (Table 2). The same geologists estimate that the younger and lower-elevation Slocum surface and associated alluvium is 400,000 to 250,000 years old (Table 2).

---

[104] The locality is north of Golden along Ralston Creek (Van Horn, 1976, p. 62-63). The "Lava Creek B" volcanic ash was formerly called the "Pearlette O" ash (see also Matthews et al., 2015).

More than anything, though, the Verdos and Slocum surfaces tell us that today's landscape in Golden was formed mostly in the last 500,000 years. If we could time-travel back to 500,000 years ago, we would easily recognize the central Golden landscape (e.g., Fig. 43).

Glacial Ice, Cannon Balls, and Paleofloods

Fig. 48. River erosion cut the V-shaped inner gorge (arrow) of Clear Creek Canyon. "M" is on Mount Zion. View west.

In the Clear Creek drainage, glaciers of the two last glacial episodes, referred to as the Bull Lake[105] and Pinedale glaciations, only reached eastward from the Continental Divide to about the intersection of I-70 and US Hwy-40 at the Empire exit, west of Idaho Springs. There, a series of moraines marks the farthest down-valley ice advance, implying that Georgetown and Empire were under ice at those times.[106] The valleys where Georgetown and Empire lie also have a typical glacial-valley U-shape, whereas the inner gorge of Clear Creek Canyon west of Golden has a distinctive river-cut V-shape (Fig. 48).

Quaternary alluvial deposits around Golden are highly modified by human activities in the last 150 years, so good outcrops are rare. Pebbly sand deposits of the Louviers and Broadway Alluvium (Fig. 46) are mostly spread along the banks of Clear Creek, 15 to 40 feet above the stream bed, around 5700 to 5800 feet elevation. A rare outcrop of Louviers Alluvium along State Hwy-58 (Fig. 49) shows the typical mixture of sediment sizes: boulder, cobbles, and sand. Many of the cobbles are rounded like cannon balls, giving the first the name of Clear Creek as "Cannon Ball" Creek by early fur trappers in the 1820s.[107] The name did not stick, though, and the creek became known as "Vasquez Fork" in 1832 and finally, Clear Creek by the late 1800s. The very youngest deposits of Recent Alluvium (Fig. 46, Table 2) were deposited over the last 13,000 years and, today, are present next to the bed of Clear Creek.

Other types of Quaternary surficial deposits are present in Golden, but their ages are not as clear. Landslide deposits are common around the edges of North and South Table Mountain (Fig. 46). They are cut up, or dissected, by erosion, having been active hundreds to thousands of years in the past, probably during wetter climate periods. More widespread are deposits called colluvium that form the "soil" in your home garden. Colluvium consists of a mixture of pebbles, gravel, sand, silt, and clay weathered directly from the underlying bedrock or deposited by wind or water. Colluvium ranges from a few inches to several feet thick.

Fig. 49. Cobbles in Louviers Alluvium along the north side of State Hwy-58 below the pedestrian bridge. View north.

---

[105] Quaternary geologists reckon that the earlier Bull Lake glaciers reached the same ending spot as the younger Pinedale glaciers. Evidence for the older glaciers was wiped out by the Pinedale glaciers.

[106] Ice covered those two towns during the advances of the Bull Lake glaciation 200,000 to 130,000 years ago and the Pinedale glaciation, 100,000 to 13,000 years ago.

[107] Thwaites, 1905, p. 279-28. Later, in 1832 French-Canadian trader, Louis Vasquez, passed through the area and renamed the creek, Vasquez Fork.

Paleofloods and the Armory Building

The Louviers and Broadway alluvium consist of glacial outwash that was brought down Clear Creek during glacial-meltwater floods. Considering the size of the large boulders in these deposits, it is hard to imagine the magnitude of flood that would pick up and carry such large material for miles out of the mountains. A well-known Quaternary geologist, Dr. Vic Baker, estimated that the discharge of a glacial flood needed to pick up and deposit the boulders in the Louviers Alluvium would be about 50,000 cubic feet per second.[108] Given that the biggest recorded flood on Clear Creek in the 20th century peaked at nearly 6000 cubic feet per second, 50,000 cfs is an almost unimaginable amount of water coming down Clear Creek. The building stone in walls of the Armory Building at 1301 Arapahoe Street (Fig. 50), derived from the Louviers and Broadway alluvium, stand as testimony to glacial flood size. Oh yes, Golden indeed rocked during those huge glacial paleofloods.

Fig. 50. Boulder walls of the Armory Building at 1301 Arapahoe Street are testimony to glacial floods that brought large rounded rocks down Clear Creek to Golden during the Ice Ages. View to east.

Completed in 1913, the Armory Building was designed by James Gow for the Colorado National Guard. As the story goes, Gow originally planned to use brick. Seeking to cut costs, he instead used boulders on the bank of Clear Creek that were left behind and nicely sorted by gold dredging operations from only a few years before (Chapter 8). Perhaps inspired by the castle-like emblem of the Colorado National Guard Engineer Company A for architectural design, Gow used 3300 wagonloads of rock[109] in constructing the building. Given his thrifty nature, one wonders if Gow used some free human-power assistance from Colorado National Guard engineering unit, Company A, made up of CSM engineering students.[110] The unique building was added to the National Register of Historic Places in 1978.[111]

From 1913 to the middle 1940s, the abundant, beautiful cobbles from the Louviers and Broadway alluvium were popular for structural walls as well as fire places and other decorative features. A version of decorative use persists today as water-tumbled landscaping material called "river rock." From the 1950s to the early 1970s, the Louviers-Broadway Aluivum were quarried sand-and-gravel for construction materials, until the quarries were exhausted and housing developments covered over new potential sites.[112]

Mammoths

Have you ever wondered why the large parking area northwest of the Morrison exit on I-70 is called the Wooly Mammoth lot? Aside from the odd spelling of "wooly," the name commemorates the fossil remains of ancient elephants that have been found around Denver.[113] For example, in downtown Golden a fossil mammoth skeleton was found in Clear Creek alluvium in 1908 during excavation for a

---

[108] Baker, 1974, and Lindsey et al., 2005.
[109] Norman, 1996, p.29 and p.63.
[110] From 1919 to 1977, all CSM first-year students were required to be in ROTC, which was held in the Armory Building (Wagenbach, 1973)
[111] Location #78000860 on the National Register of Historic Places in the U.S. (US Dept of Interior, 1992)
[112] Arbrogast et al., 2002, p. 27. Also discussed in Chapter 11.
[113] As per Kermit Shields, these were more likely Columbian Mammoths because the range of Woolly Mammoths was farther north.

sewer line. Previously in 1872, Edward Berthoud found an "elephant" (mammoth) tusk in Clear Creek alluvium perched in a gravel deposit about 25 feet above the creek bed near Guy Gulch, 5 miles west of Golden.[114] Sewer excavations in Denver uncovered other finds of camel, bison, and horse teeth and bones.[115] During cold glacial-climate periods, mammoths and other large mammals roamed the landscape around Golden, a startling contrast to today's fauna.

Channeling Clear Creek

Before about 1870,[116] the Clear Creek channel through central Golden was wide, with sand bars intertwined like braids (Fig. 51). A braided channel pattern is common where a stream channel changes from a confined zone, as in the inner gorge of Clear Creek, to a broader unconfined area. As streamflow

Fig. 51. 1882 view of Golden showing the braided channel (arrow) of Clear Creek. Smoke in background is from a smelter. View east. X9791, DPL WHC, used with permission.

spreads out at the mouth of a canyon, the water cannot carry as much sediment, so the sediment drops out, or is deposited, commonly into a series of braided sand and gravel bars.

Since the 1880s, Clear Creek channel has been greatly modified. First modified in the late 1870s and 1880s, Coors Brewery built large ice ponds on the south side of the channel, east of Washington Street. Gold dredging from 1904 to 1908 further modified the Creek channel east of Coors Brewery leaving behind large piles of boulders and cobbles from gold dredging operations.[117] Eventually Clear Creek was completely straightened to become a flood-control channel through the Clear Creek Valley

---

[114] Berthoud, 1872 and Rockwell, 1872, p. 302
[115] J.H. Johnson, 1930, p. 376.
[116] An 1878 city map of Golden (Willits, 1878) also shows the braided pattern of sand and gravel bars in the reach from the mouth of Clear Creek to Washington Street, preceding upstream hydraulic mining operations.
[117] Placer gold operations in the Golden area are discussed in Chapter 8.

east of the Ford Street Bridge.[118] West of Ford Street, human activity gradually modified the original sand bars (Fig. 51) of Clear Creek. In 1886, hydraulic gold mining near the canyon mouth filled-in the channel with sediment from those operations (Chapter 8). The City of Golden built its water-supply ponds along the channel on the north side of the Creek in the late 1800s. The Creek channel was further filled in to create the western end of 11th Street around the time that the Mines Experimental Plant was built in 1912. When US Hwy-6 was constructed at the canyon mouth in 1950, bridging Clear Creek required narrowing the channel below the highway. At the time, like many rivers in the U.S., Clear Creek was viewed as more of a flood-control channel (and an open sewer) rather than an amenity that people might enjoy. However, in 1998, the channel was restored for recreation by creating bends and pools within the channel between the US Hwy-6 and Ford Street bridges, forming the now-very popular kayak run and open-space trails on either side of the Creek.

Historic Flows and Floods

Flows in Clear Creek are low for much of the year, less than 250 cubic feet per second (cfs), making waders and tubers happy. In wet years, spring runoff from snowmelt can reach up to 2000 cfs, making kayakers (skilled ones) happy.

With Clear Creek essentially un-dammed upstream, floods historically have caused major havoc in Golden. As we all know, June through September (and rarely May) can bring some intense rainstorms that result in damaging floods, as represented by peak flows over 3000 cfs since 1911[119] (Fig. 52). The

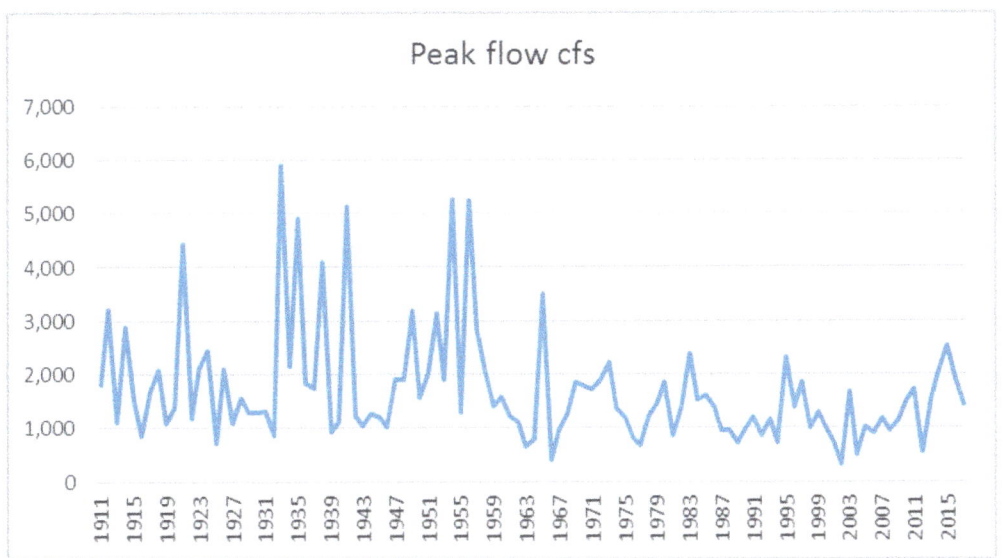

Fig. 52. Recorded annual peak flows on Clear Creek gauges near Golden, 1911-2017. Data from USGS gauges #06719500 (1911-1974) and #06719505 (1975-2017).

largest flood on Clear Creek in the recent history of Golden probably occurred on 24-25 July 1896. The peak discharge in the neighboring Bear Creek drainage to the south was an estimated 8600 cfs.[120] Based on the damage in Golden and a Clear Creek watershed 67% greater than that of Bear Creek, the flows on Clear Creek in Golden were likely higher!

---

[118] Arbrogast et al., 2002.
[119] The first stream gauge near Golden was installed in 1911, upstream from the Church Ditch headgate. The current gauge at the US Hwy-6 bridge was installed in 1975. The data from the two gauges are compatible, as there are no tributaries entering Clear Creek between the site of the former 1911 gauge and today's gauge.
[120] Mile High Flood District, 2020 https://www.udfcd.org/FWP/ebb/bear_history.html .

The highest recorded peak discharge on a Clear Creek gauging station near Golden was 5890 cfs on 9 September 1933.[121] At these highest flow rates, the large cobbles on the bottom of the channel rattle around noisily, but even at these flow sizes, cobbles do not move downstream very much, if at all.

The four Golden tributaries to Clear Creek, Cressman/Tucker Gulch, Chimney Gulch, Deadman/Kinney Run Gulch,[122] and Apex/Lena Gulch,[123] have also experienced damaging floods when intense rainstorms hit in just the right spots. For example, a wall of water coming down Tucker Gulch in the 1896 flood, killed three people.[124] An 1894 flood inundated Kinney Run Gulch, across the area of today's Coors Brewing (Fig. 53). The September 2013 floods that wreaked havoc from Boulder to Fort Collins registered a mere 1020 cfs at peak flow on the Clear Creek gauge in Golden. However, Apex Gulch sustained major damage because an early part of the storm system dropped buckets of water in just the right spot. In the narrow confines of Apex Canyon, water and debris flows completely washed out and/or buried hiking/biking trails. The lesson: it does not take much concentrated rainfall in just the right spot to make a mess.

Fig. 53. Clean-up after an 1894 flood affecting the confluence of Kinney Run Gulch with Clear Creek in Golden. Coors home (on site of current Coors Brewery) on right. Golden smelter in background on north side of Clear Creek. View northeast to North Table Mountain. X-63151 DPL WHC, used with permission.

---

[121] USGS stream gauge data for previous gauge #06719500 upstream from today's gauge. Curiously, Follansbee and Sawyer, 1948, incorrectly report a large flood in Golden in August 1888. It was only a local flash-flood near Georgetown, barely being mentioned in the CT 8 August 1888.
[122] The middle part of Kinney Run Gulch is called "Deadman Gulch" on the current (2016) USGS topographic map of the Morrison Quadrangle. We use "Kinney Run Gulch" for the entire length because of the city-designated Kinney Run Trail that runs along it. Also, the Willits 1878 map of the City of Golden shows it as Kinney Run.
[123] Lena Gulch was called Coon Gulch on the 1888 and 1899 USGS topographic maps of Denver and called Apex Gulch on the Willits 1899 Farm map. The name changed to Lena Gulch on the 1938 USGS topographic map.
[124] CT 29 August 1896.

Chapter 7: Water is Golden

Living in Golden and the greater Denver area now with all the green trees and lawns, we forget that we are actually living in a semi-desert. On 5 July 1820, Stephen H. Long and his men camped for few nights at what is now the confluence of Clear Creek and the South Platte River.[125] The next day four men in the party headed up Clear Creek, intending to walk all the way to the western foothills in Golden. They described the surrounding plain as covered with prickly pears (cactus) and "sparse rigid grasses with barbed seeds like the quills of a porcupine."[126] Cactus and prairie grasses (Fig. 54), not lawns and trees, are the native vegetation of the semi-arid region surrounding the greater Denver area, including Golden.

Fig. 54. It is hard to believe that the heavy growth of trees in Golden today masks the reality of Golden's semi-arid climate depicted circa 1872. What a difference irrigation water makes. View southeast at Castle Rock. X-10006 DPL WHC, used with permission.

When Euro-Americans came to Golden in 1859, they needed water for drinking, cooking, and livestock-watering, as well as to establish farms to grow food. Farms required water in the form of irrigation, and perennial creeks like Clear Creek were steady, sure water sources. Little wonder that settlers established western towns like Golden along perennial streams and rivers. Early Euro-American settlers not living next to perennial streams commonly used shallow, hand-dug groundwater wells as a water source: a lot of work! Groundwater is still used in Golden, but not commonly for drinking water, because the Golden municipal water system serves most homes and businesses.

Resourceful Clear Creek

Clear Creek is the main source of drinking and irrigation (now mostly for landscaping) water in Golden. It is also a main source for downstream users such as Coors Brewery for industrial use, and for the cities of Arvada, Thornton, and Westminster. Today it provides drinking water for more than 300,000 people.

Euro-American settlers in this semi-desert region diverted water from the Creek for various uses. The first irrigation diversion in northeastern Colorado[127] was right in Golden. In April 1859 David K.

---

[125] Thwaites, 1905.
[126] Ibid. p. 281.
[127] However, irrigation in the San Luis Valley in southeastern Colorado began in 1852, and Costilla County was awarded Priority No. 1 in Water District 24 of Water Division 3 dated 10 April 10 1852 (Colorado Water Institute, 1952).

Wall diverted water from Tucker Gulch for his vegetable farm near the current railroad yards in east Golden. This was also Golden's first vegetable garden. Wall made more money selling vegetables than did most miners seeking gold.[128] The first two permanent diversion canals, or ditches, began construction in late 1859, finishing in early 1860. The diverted water operated sluiceboxes for mining placer gold in the gravels at Arapahoe Bar, located three miles east of downtown Golden (see Chapter 8, also). It was one of the first gold discovery sites along Clear Creek. Later these two ditches became the Wannemaker-Arapahoe and Farmers Highline Ditches, used for irrigation.[129]

After 1860, surface water was diverted from Clear Creek for Golden and other irrigation users through ditches that started in and upstream from Golden. The first two irrigation ditches completed were the Rocky Mountain (1862) and Church (1865) Ditches. The Welch (1871, see below) and Agricultural (1874) Ditches were built afterwards. The timing is important due to water rights, discussed below. All or portions of these ditches are still used. Modern use, however, is mostly for domestic and industrial purposes, rather than agricultural irrigation. How much any user can take, and within which part of a year, is the subject of water rights.

Whiskey is Made for Drinking

Water in the Western U.S. is a scarce resource that is shared by virtue of water laws.[130] Water rights to perennial streams and rivers are a property right, which means they can be bought and sold in pieces or wholesale. To own a water right, one must have a water source, a beneficial use, and exercise that use (known as the "Can and Will" doctrine). With a scarcity of available water, water rights are dearly sought and held, as you might imagine. Also, when water rights and diversion amounts were defined in the 1860s and 1870s, data did not exist to understand typical streamflows, especially during drought years. As a result, more water has been assigned to users than is available from Clear Creek during drought years. Known as "over-appropriation," this situation is common with most streams and rivers in the western U.S. It makes for a lot of legal wrangling!

Golden, like many users in the Western U.S., derives its water rights under the "First Use" doctrine[131] dating from 1861 in Colorado Water Law. Golden has senior rights from October thru March (the non-growing season) and junior rights in the growing season. During the spring and summer months (growing season), other users have a higher priority, including those users that have purchased historical agricultural water rights. The City of Golden maintains a projection of the drought impact on City water supply relative to its water sources and senior and junior rights (Fig. 55). To date during drought years, such as that in 2002, Golden has imposed only voluntary water restrictions, because it has large water sources and more-senior rights relative to other municipalities. However, understanding those rights and protecting them comes under the heading of "It's Complicated." One takeaway from Figure 55 is how important Guanella Reservoir, west of Empire, is to maintaining Golden's water supply. The reservoir was built after the 2002 drought-year, to ensure a reliable supply for city residents.

You may be surprised to know that Golden also has a special water right for its Kayak Course along Clear Creek, called a "Recreational In-Channel Diversion." This right maintains the flow of water through the Kayak Course during the summer months, which keeps tubers and waders happy. Immediately challenged in court, this Kayak Course water-right was a precedent-setting case: the first of its kind in the Western U.S.[132] The right was upheld in court and is now an accepted part of Colorado water law.[133]

---

[128] Gardner, 2020, p. 24. Wall was also a founder of Golden when formed on 16 June 1859.
[129] Western Mountaineer, 7 December 1859.
[130] Mark Twain allegedly said, "Whiskey is for drinking; water is for fighting over." Even though he may not have made the statement, it is a true witticism.
[131] In the western U.S. the first person/organization to claim a water right gets the first opportunity to use it, which is also referred to as a "senior" water right. This is also known as the "first in time, first in line" or the prior appropriation doctrine.
[132] Benson, 2016, https://scholarship.law.berkeley.edu/elq/vol42/iss4/1/
[133] Ibid. p. 753.

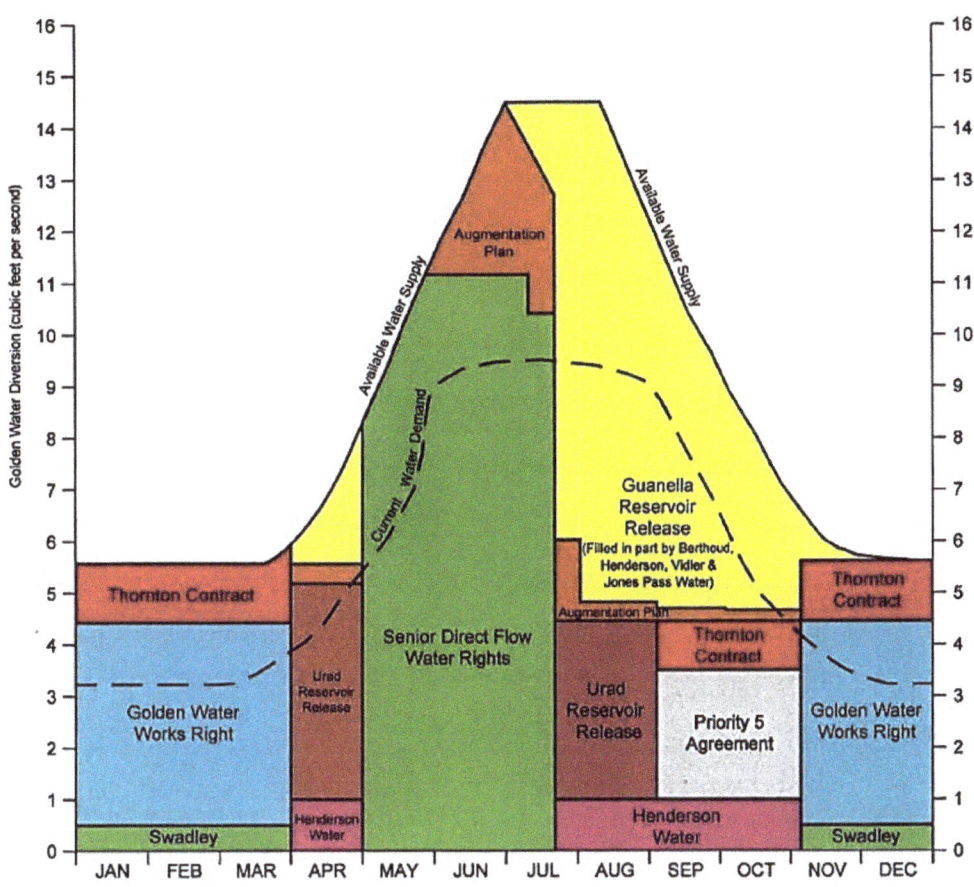

Fig. 55. Golden water rights, supply, and demand based on extreme drought conditions for estimation of water-rights yields. Courtesy of the City of Golden Water Department, updated for 2021.

Golden's main water rights are of three types: Direct Flow, Release from Storage, and Trans-Basin Transfers. Direct flow is obviously from Clear Creek. Storage releases are from the Guanella and Urad Reservoirs near Empire, which are within the Clear Creek watershed. Golden's trans-drainage basin transfers include water from Peru Creek via the Vidler Tunnel under Argentine Pass near Georgetown, which derives water from a different watershed, in this case from the other side of the Continental Divide.[134] Interestingly, the water from those storage and transfer sources does not come directly to Golden. Rather Golden releases carefully measured amounts of water from these sources into Clear Creek higher up in the mountains. The water travels down Clear Creek to the City of Golden headgate where Golden then diverts the same amounts of water as that released in the mountains. Golden's headgate is about 20 feet upstream of the the Church Ditch headgate, both about ½ mile upstream from US Hwy-6. The water enters a pipeline under the Grant-Terry Trail and travels to the settling ponds west of the Golden Water Works. All the water transferred from streams and canals (ditches) is carefully measured by stream gauges in order to comply with the assigned water rights.

---

[134] See https://www.arcgis.com/apps/MapJournal/index.html?appid=f4c95c11c955458d85b1d1934b137871 a City of Golden, GIS story-map discussing Golden's water sources.

Historic Welch Ditch

Of the ditches in the Golden area, Welch Ditch is one of the most historic, due to its wooden flume that ran from near the headgate west of Tunnel 1 (Fig. 56) to Golden along the south side of Clear Creek Canyon. Charles Welch built the first segment in 1871 to provide water to the Colorado School of Mines at its original location at the south end of today's Ford Street.[135] Like all irrigation systems before the use of electrical pumps, Welch Ditch was a gravity system designed to maintain a non-erosive low-velocity current, which mandated that it follow a winding path along topographic (elevation) contours. The ditch wound around and crossed today's US Hwy-6, crossed the Rubey Mine, and then went south through the CSM campus, reaching its southern end at the first site of the School of Mines (Fig. 56). In 1873, partners Charles Welch (1830-1908), Edward Berthoud (1828-1910), and William H. Loveland (1826-1894) extended Welch Ditch from the then-Territorial School of Mines site back to the north,

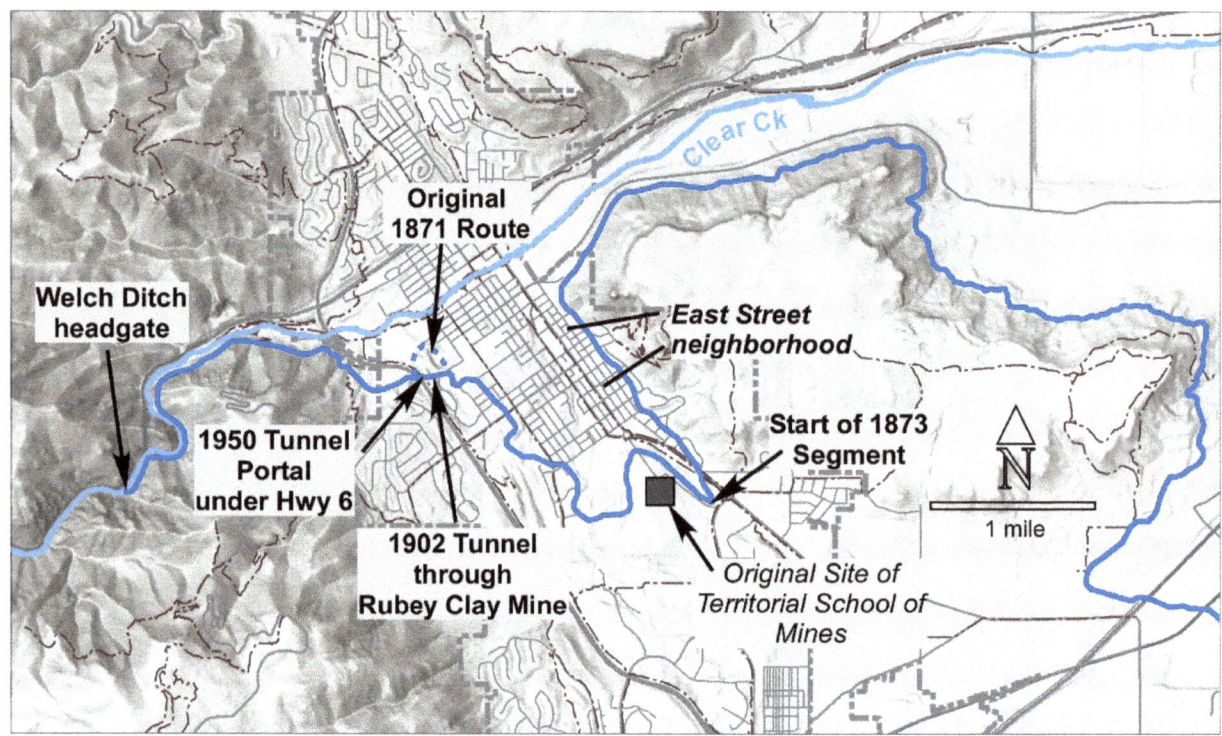

Fig. 56. Welch Ditch winds eastward through and across Golden, turning sharply at the 1873 site of the Territorial School of Mines, now the Lookout Mountain Youth Services Center. It went north through the East Street neighborhood.

traversing around the west, north, and east sides of South Table Mountain. The extension provided water to farmers in today's East Street neighborhood and for Welch's and Loveland's lands east of the mountain.[136]

Originally, Welch Ditch wound around the Laramie outcrops at the north end of the Rubey Mine near today's 12th Street. But the path was altered in 1902 (Fig. 56). A cutoff tunnel was built directly through the vertical Laramie sandstone beds of the Rubey Mine, permitting more clay mining and eliminating leakage into the mining operation. About same time, G.W. Parfet successfully sought a new right-of-way from the City of Golden to expand the mine west of the then-smaller CSM campus. Even so, the tunnel through the sandstone ridges impeded clay mining. Apparently mine cars could not be

---

[135] The site of the current Lookout Mountain Youth Services Center, originally called the State Industrial School.

[136] With the Welch Ditch extension to east side of South Table Mountain completed in 1885, C.C. Welch and Loveland formed the Lakewood Company and platted 83 acres along Colfax Avenue in 1889, leading to their founding of Lakewood. Welch's "country" home was his Welchester Estate, where Welchester Elementary is now located east of Colorado Mills Shopping Center on 10th Avenue.

Fig. 57. Abandoned Welch Ditch tunnel portal on east side of U.S. Highway 6. Location shown in Fig. 56.

heaped very high in order to get under the ditch tunnel, much to the aggravation of Parfet.[137] In 1950, a siphon-tunnel was built under US Hwy-6 to connect the surface ditch to the 1902 cutoff tunnel in the Rubey Mine on the east side of the highway (Fig. 56). The siphon tunnel provided some excitement for daring (and crazy) teenagers who would "surf" the waters during the summer season.[138] The siphon-tunnel and portal into the now-abandoned Rubey Mine are now closed off (Fig. 57)

Use of Welch Ditch in Golden ended in 2002 with construction of Fossil Trace Golf Course, which opened in 2003. Years ago, the Agricultural Ditch Company purchased Welch Ditch water rights, today diverting that water into its Ditch at the headgate on Clear Creek east of Ford Street, leaving the old Welch Ditch mostly high and dry. However, stretches of Welch Ditch continue to function for drainage throughout the City of Golden.

Now an official historical feature, the westernmost part of Welch Ditch has become part of the Clear Creek Park Open Space and trail system administered by Jefferson and Clear Creek Counties as part of the Colorado Peaks to Plains Trail System. The former Welch Ditch wooden flume is now restored as the "Welch Ditch" hiking trail as part of the Peak to Plains Trail project, administered by Jeffco Open Space.

Tales of Metals and Nutrients

In the United States we take for granted that our drinking water is clean and safe to drink out of the tap. Inasmuch as Golden's municipal water comes straight out of Clear Creek, the quality, or chemical composition, of Clear Creek water is a critical issue in how the City of Golden treats its municipal water, and what it costs to treat it.[139] And, because water keeps flowing downstream, what gets into the Creek below the Golden headgate is also important. Simply put, tubers, waders, kayakers, and fly-fishermen need good water quality in Clear Creek for those activities. Downstream users, like Coors Brewing, the ditch companies, and the cities of Arvada, Thornton, and Westminster require good water, too.

Like water laws governing how much water users can take out of Clear Creek at any given time, laws regulate the quality of that water. The Clean Water Act of 1972 sets standards for surface-water chemistry, among other things, and mandates that point-source[140] wastewater dischargers renew their permits every five years. That renewal process keeps treatment up to date with improving technology and changing water-quality standards. That renewal process has been a big driver for cleaning up pollution in Clear Creek and other waterways in the U.S.

Up to 1/3 of your monthly water bill is for treating the water that goes down your drains and into the city sewer system. That water is collected then treated at a Wastewater Treatment Plant on 44th Avenue next to Clear Creek, operated by Coors Brewery. Before being returned to Clear Creek, yours

---

[137] Parfet, 2007.
[138] Chip Parfet recounted this in an interview in 2020. He lived to talk about it!
[139] Golden conducts annual water quality tests and sends results to its water users. Annual reports are available at: https://www.cityofgolden.net/government/departments-divisions/water/drinking-water/
[140] As per the Clean Water Act of 1972, point-sources are "any discernable, confined and discrete conveyance…from which pollutants are or may be discharged." Two common point-source dischargers are sewage treatment plants and factories, including breweries.

and others' wastewater is treated to the legal standards mandated by the operating permit from the Colorado Department of Public Health and Environment (CDPHE). Treatment costs money. Nothing is free.

The water quality of Clear Creek upstream of Golden has been a discussion topic since the late 1800s when irrigation-ditch users complained about ditch water killing their crops, and Golden residents complained about Clear Creek water corroding their pipes. Since the 1970s, study and cleanup became intensive, under regulation by the U.S. Environmental Protection Agency (EPA) and the CDPHE. With a huge diversity of users, sometimes with competing interests, the Upper Clear Creek Watershed Association was formed in 1993 to foster cooperation among water users.[141] The modern story of water quality in Clear Creek is a tale of two pollutants: metals and nutrients.

Orange Water: Metals

In May 1980 a slug of orange-colored water surged out of the old Argo Tunnel in Idaho Springs (Fig. 58) and made its way down Clear Creek.[142] Although judged not to be a human health hazard because it was quickly diluted by the high spring runoff in the Creek, the orange blob was a wake-up call to downstream users. By the 1980s enough data had been collected at various stream gauging stations along Clear Creek, including the gauge at Golden, to show that water quality in Clear Creek was severely degraded with heavy metals, especially lead, cadmium, iron, zinc and copper, from the release of acid water from mine tailings and mine shafts in the Idaho Springs-Central City area upstream from Golden. As shown by data from the sampling point on Clear Creek at Golden (now sampled annually since 1994), copper, zinc, and cadmium greatly exceeded their standards, and iron and manganese amounts were at their legal limits in 1987.[143] These issues led to high costs for treating Golden's drinking water for safe consumption. It also created poor habitat conditions for fish life within Clear Creek along its entire length.

Fig. 58. Argo Tunnel portal in 2004 before remediation. Orange water is laden with heavy metals that, since 1998 have been removed before discharging to Clear Creek. 5401076.jpg, DPL WHC, used with permission.

In 1983 the EPA identified the Idaho Springs-Central City area as a Superfund Study Area, and proceeded, in partnership with other agencies like the CDPHE, to identify specific localities for cleanup that would make a big improvement in Clear Creek water quality. By 2006 acid mine-water remediation activities in the Clear Creek watershed, such as the Argo Tunnel remediation project in 1998, had achieved significant metal-load reductions, and stream-standard exceedances became relatively infrequent.[144] As testimony to the success of the cleanup efforts, the West Denver Trout Unlimited organization made habitat improvements in Clear Creek for a mile upstream (the Golden Mile) of the Golden Kayak Run in 2009. Now trout live in the Creek, making fly-fishermen happy. For municipal water users like the City of Golden, reduction in manganese in particular was an improvement, because manganese is expensive to remove from the water. If left in the water, it stains laundry and causes poor water taste. For many years now, water quality for Golden water meets or exceeds all Federal and State requirements.

---

[141] Upper Clear Creek Watershed Association, 1997.
[142] Constructed in 1893 to drain acid mine-water from the network of mine workings, the Argo Tunnel emptied directly into Clear Creek until the EPA constructed a water treatment plant to clean up tunnel discharge in 1998.
[143] Tudor Engineering, 1987, p. B-2.
[144] Upper Clear Creek Watershed Association, 2014, p. E-2.

A Quiet Legacy

Did you know that, before remediation, there was an EPA Superfund Site right next to Clear Creek, on the north side of the CSM campus? It is a little known and discussed story with its origin in 1912, when the CSM Experimental Plant (also see Chapter 8) was built along the bluffs along the south side of Clear Creek (Fig. 59). Operating from 1912 to 1987, it focused on processing metallic, non-metallic, and radioactive ores.[145] Initially, through the 1940s, waste was discharged directly to Clear Creek. Later facilities included a tailings pond that stored a mix of fine solids and liquids. The pond was located on the south bank of Clear Creek, across from today's Lions Park baseball fields. The pond materials were contaminated, as was the dirt around and within the facility, and all of the drains and sewers leading from the buildings. Remediation for pollution began immediately when the facility shut down in 1987. By 1989, the area was declared safe, and a probability of erosion of the former tailings pond next to Clear Creek by natural flooding was considered nil.

Fig. 59. Experimental Plant building shortly before demolition in 1997 and before construction of the Golden Kayak Run in 1998. View to southwest across Clear Creek, in foreground. Barbara Warden photo, used with permission.

As usual, there is a <u>however</u> in the story. In the remaining buildings, a water pipe broke in January 1992 (frozen pipes, anybody?), flushing the drains around the facility and releasing thousands of gallons of radioactive nuclides and heavy metals directly to Clear Creek, contaminating the water for over 250,000[146] downstream consumers. Not good. The EPA stepped in and initiated a CERCLA response action.[147] More remediation from 1992-1997 cleaned up the buildings and the drains. When the buildings were finally torn down in 1997, the land surface was capped with clay to prevent movement of contaminated dirt and leached material from the site. The former building site basically became a landfill. Drilling permanent groundwater-monitoring wells near Clear Creek ensured that underground contaminants were not reaching Clear Creek. In 2014, during construction of a new soccer field, engineers removed the contaminated dirt from the landfill to an approved low-level waste facility outside of Golden, ensuring that no hazard remained. At the same time, engineers remediated the old tailings pond along Clear Creek. Now, the popular area is a soccer field, parking lot, and part of a paved hiking/biking trail along the south side of Clear Creek.

Modern Slime: Nutrients

By 1990, Standley Lake was dying. It is not in Golden, so why care? Well, because everything is interconnected, Standley Lake <u>was</u> a problem to Golden because part of the water that feeds into the lake comes out of the Church Ditch headgate, about 20 feet downstream from where Golden takes its water. Standley Lake was dying because the water from Clear Creek was overloaded with nutrients like nitrogen and phosphate, the ingredients of fertilizer. The nutrients came from upstream in the Clear Creek watershed from wastewater treated to low standards. Those nutrients caused algal blooms in the lake: essentially pond scum. They also contaminated Standley Lake water, which serves the residents of Thornton, Northglenn, and Westminster. Gross.

---

[145] Oredigger, November 2016, v. 97, Issue 4: https://oredigger.net/2016/11/uncovering-the-little-known-history-of-mines-research-institute/ is the source for this discussion.
[146] This was the population served by Clear Creek downstream from Golden in 1992.
[147] CERCLA means Comprehensive Environmental Response, Compensation, and Liability Act, called "Superfund" for short.

In 1991 gambling casinos and hotels came to Black Hawk and Central City as a result of popular vote by Colorado citizens. With that development came the inevitable stream of wastewater from more visitors. The pre-development wastewater treatment facilities, built for much lower capacity, became overwhelmed. After much argument and a cease-and-desist order, a new treatment plant in Black Hawk was finished in 2005 to eliminate that wastewater pollution to Clear Creek.

The nutrient issue is a big reason why it costs money to treat wastewater, like you see on your water bill. Not just the upstream users have to pay attention to the law. We, who live downstream, use that water, returning it back to the stream after treating it ourselves. Like rocks, water has a cycle, too. Everything is interconnected.

The Future

As more and more point-sources are cleaned up by regular discharge-permit renewal, those easy-to-find sources go away. The difficulty then becomes pollution sources that are spread out across large areas and are not easy to pinpoint. These are called "non-point" sources. And, guess what? They include you and me.

Fig. 60. No magic! Everything drains to Clear Creek. Emblem on storm drains around Golden.

One of the biggest and difficult-to-control non-point sources is runoff water from highways, streets, and even your driveway. Ever see the little circular signs (Fig. 60) on the curb above your neighborhood storm drain that says "No Dumping, Drains to Creek?" Every time you wash the yucky guck off your car in your driveway, that stuff eventually flows into Clear Creek. That same guck also drains off the highways during storm runoff, which is why the CDPHE requires modern highway design to include storm-water detention basins, such as the new one off US Hwy-6 in the lower Chimney Gulch Trail area.

Yet another non-point challenge is spillage from traffic accidents in Clear Creek Canyon. While these cannot be prevented, quick communication by first responders lets downstream users like Golden shut the water-intake portals until the slug of pollution has passed by. Good communication is obviously paramount and fostered by the collaboration of emergency responders and upstream users. Other non-point sources in the Clear Creek watershed are the many private septic systems in the Front Range west of Golden. Little by little several counties have tightened the regulations for private septic systems, requiring more frequent inspections to ensure the systems work properly.

Water Below the Ground

Groundwater exists below the land surface in the tiny holes, or pores, between sediment particles, and in the tiny cracks, or fractures, in hard igneous and metamorphic rock, like basalt, granite, and gneiss. As you might imagine, the amount of water stored in the tiny holes is a lot less than what can be stored in a pond or reservoir. In addition, how fast groundwater can flow through the tortuous paths between tiny holes is a lot slower than flow in a creek or river. But, the volume of rock that contains groundwater, an aquifer, is so huge that the quantity of groundwater is enormous.

Groundwater is a key aspect of the water cycle.[148] In everyday experience, we mostly deal with shallow groundwater, or groundwater that is in reach of wells less than 100 feet deep. Groundwater comes from rainfall, snowmelt, and in some areas, human-built facilities like septic systems and unlined canals or ditches, in which water seeps directly into the ground. Importantly, groundwater and surface water are interconnected, although different water laws govern the two.[149] Let it be said that for a long time, people considered the two types of water to be "different." That led to some users pumping water out of the ground right next to a river, thinking that they were not really using river water. Untrue. For

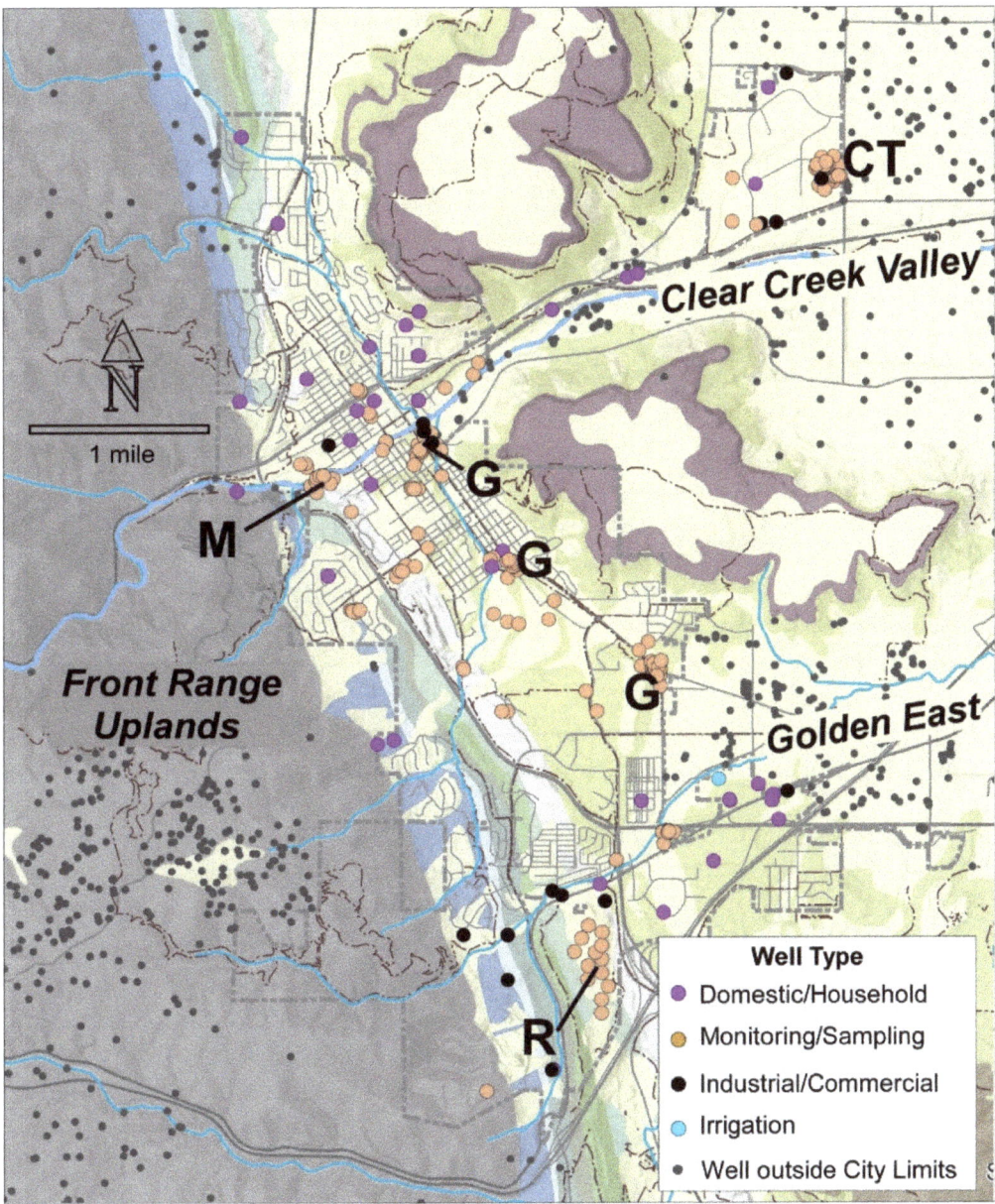

Fig. 61. Underground aquifers are basically the same as rock formations. Hundreds of groundwater wells exist in the greater Golden area. R=Rooney Road monitoring area, M=CSMRI monitoring area, G=monitoring around former gasoline stations, CT= Coors Technology Center. Well data from State of Colorado, Division of Water Resources (2019). Geology as in Fig. 4.

---

[148] See https://www.usgs.gov/special-topic/water-science-school for a good primer on water and the water cycle.
[149] In Colorado both are administered by the Colorado Division of Water Resources in the Department of Natural Resources. http://www.water.state.co.us/Home/Pages/default.aspx

example, when some users of Platte River water found out that their water was reduced by pumping, they went to court, and the pumping was stopped. Ground and surface water are connected.

In the Denver Basin and Front Range, including Golden, rock formations discussed in Chapters 2, 3, 4, and 6 are grouped into different aquifers (Fig. 61). [150] For any given water well, you are only allowed to drill deep enough to take water from one of the legal aquifers (as we said, there are rules). The shallowest aquifers are those of Quaternary alluvium, especially the Louviers, Broadway and Recent alluvium, which are concentrated along Clear Creek Valley within Golden (Chapter 6). Next-most common and important are the layers of the Denver, Arapahoe, Laramie, and Fox Hills formations (Chapter 4). These formations make up the major groundwater aquifers of northeastern Colorado. Areas of older sedimentary bedrock, like rocks of the Dakota Group (Chapter 3), and the Lyons Formation (Chapter 2) are also aquifers, but they are very deep and generally not used for groundwater supply. Precambrian metamorphic rocks (Chapter 2) make up the most important aquifer in the the Front Range uplands (Fig. 61). You may rightfully wonder how Precambrian gneiss can be an aquifer. In those rocks, groundwater is stored in and flows through tiny interconnected and numerous fractures. Fractured-rock aquifers are present throughout the towns and housing developments across the mountainous Front Range.

Golden Water Wells

Over 250 groundwater wells exist within the Golden city limits (Fig. 61). Most of the wells reach less than 60 feet below the ground surface.[151] Monitoring wells, used to track water levels and groundwater quality constitute almost 200 of the groundwater wells within the city limits. One cluster in the area of the Rooney Road Athletic Complex ("R" on Fig. 61) monitors groundwater around the former Jefferson County sanitary landfill (Chapter 10). Another cluster east of the mouth of Clear Creek Canyon ("M" on Fig. 61) monitors groundwater quality in association with the former CSMRI (CSM Experimental Plant) site. Clusters in central Golden monitor water quality near former gasoline stations that had leaky underground tanks. The cluster at the Coors Technology Center ("CT" on Fig. 61) monitors industrial pollution plumes. Monitoring wells are usually flush with the ground surface and capped with a small-diameter marker (Fig. 62).

Fewer than 50 wells serve other purposes in Golden. Less than 20 wells are used for domestic or household supply in the city limits (Fig. 61). The number is small because the municipal surface-water supply system, drawn from Clear Creek serves most city residents. Commercial and industrial wells can reach up to hundreds of feet deep. Seven such wells in south Golden supply industrial make-up water for businesses along US Hwy-40/Colfax and to the Martin Marietta Spec-Agg Quarry near the former Heritage Square area. Three industrial wells in central Golden supply make-up water to Coors Technology Center and Coors Brewery.

Outside the city limits, most groundwater wells are for domestic/household use. They typically have discharge rates less than ten gallons per hour. Such wells in the "Golden East" area (Fig. 61) are usually over 100 feet deep and draw water from the Denver Formation. Domestic wells in the "Front Range Uplands" (Fig. 61) take groundwater from Precambrian fractured gneiss aquifers. Well-depths in the mountains vary from tens to several hundred feet below ground

Fig. 62. A groundwater monitoring well along South Golden Road, near Lookout Mountain Youth Services Center and Golden High School.

---

[150] See https://dwr.colorado.gov/services/well-permitting/denver-basin
[151] State Department of Water Resources water-well database, 2019.

surface. In the Clear Creek Valley area east of Golden, wells include a mix of sampling/monitoring, industrial and domestic-supply wells. Wells directly along Clear Creek draw from Quaternary alluvium, whereas north and south of the Creek they draw from the Denver Formation.

Springing Forth

A spring is where groundwater comes to the land surface, creating a wet area like a marsh, pond, or even a stream. A perennial spring runs all year whereas an intermittent spring runs only during the wet season. Most people think of springs as "natural" features in the landscape, feeding water to a pool of surface water, like the spring-fed Soda Lakes northeast of the intersection of C-470 and US Hwy-285 southeast of Morrison. Seasonal or intermittent springs are fed by snowmelt and form temporary ponds in the spring and early summer. They dry up when the water source dries up. Other types of intermittent springs may form under your house during wet years when snowmelt seeps into the shallow ground. In the middle 1800s when Golden was settled, only a few natural springs existed in the area. In reality, the 1869 Hayden Survey encountered[152] few if any permanent springs across the entire Great Plains as a result of the arid climate. Climate change toward drier years makes intermittant springs rarer and rarer.

Human-made springs can be created by the presence of an unlined irrigation ditch that leaks water into the ground. Seasonally-applied water to lawns and other landscaping can also create human-made springs. Groundwater pumping can extinguish a spring by diverting the water. If pumping stops, the spring can reappear when groundwater levels rise high enough to come up to the land surface again.

Springs reported in the Golden area are concentrated on and around North and South Table Mountains (Fig. 63).[153] On South Table Mountain, seven intermittent springs occur at the base of the basalt caprock. Surface water seeps down into fractures in the basalt where it becomes groundwater.

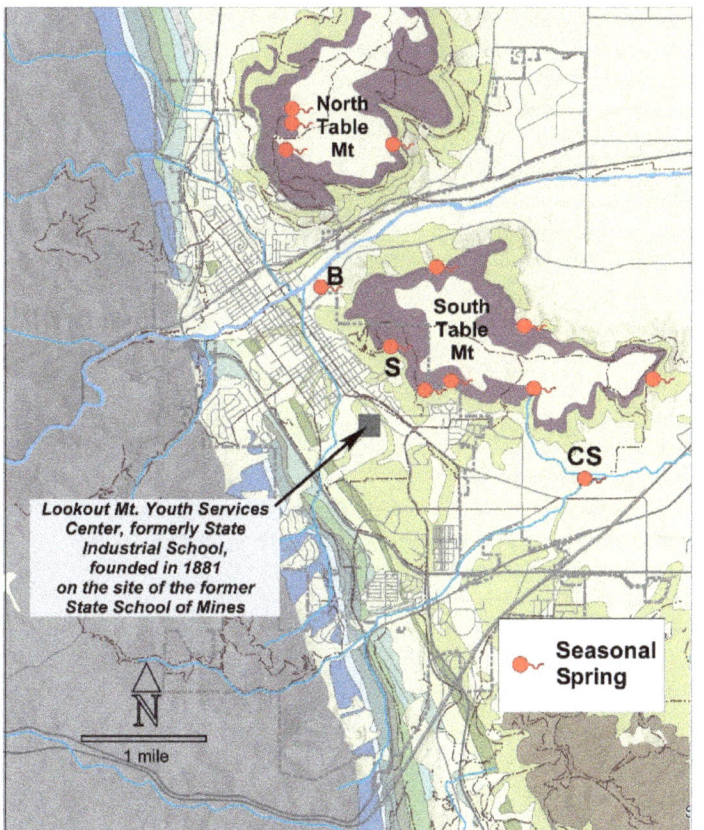

Moving underground and downslope toward the edges of the mountain, the seepage, or groundwater, emerges as seasonal springs near the base of the basalt caprock. The areas around the springs have lush vegetation because they are wet throughout spring and into the late summer.

A spring on the west side of South Table Mountain was, for a while, the source of drinking water for the State Industrial School (now the Lookout Mountain Youth Services Center). The site of the spring is likely above the Table Heights neighborhood on the west side of South Table Mountain (location "S" on Fig. 63), where seasonal discharge is still common. When the State

Fig. 63. Springs in the Golden area. B=Coors Brewery, S= 1881 Spring for State Industrial School, CS=Cold Springs. Compiled from USGS topographic maps, Rousseau (1980), Willits 1899 Farm Map, old aerial photographs, written and oral reports.

---

[152] Hayden, 1869, pgs. 10 and 37. Notable exceptions include spring-fed Soda Lakes south of Bear Creek and, of course, the area we now know as Colorado Springs.
[153] Compiled from the Golden and Morrison 1:24000-scale USGS topographic quadrangles, from old photographs and historical references as discussed in text.

Industrial School opened in 1881 on the former site of CSM, two shallow groundwater wells existed on the property, but they supplied insufficient water. In 1881, a spring "about 4300 feet to the east," on the west side of South Table Mountain, was developed by enlarging the area around the spring, sinking a 15-foot-deep well at the site as a reservoir, and laying 1¼-inch diameter galvanized iron pipe over to the school. The spring initially yielded 2000 gallons a day of the "purest water."[154] By 1900, spring discharge was of such low volume that Saturday "bath night" at the School required lots of planning.[155] Thus, in 1905 a large, new water well was constructed at the school site to supply water.[156]

Another historic spring reported by early settlers and shown on 1800-vintage maps is Cold Spring ("CS" on Fig. 63), along Lena Gulch, east of Golden. That spring reportedly had excellent, cold water, leading to establishment of a ranch by that name in 1860. The ranch and spring had a strategic location along the road to the gold fields.[157]

Four seasonal springs on North Table Mountain are also associated with the basalt caprock. Groundwater seeping from three springs on top of the mountain (Fig. 63) keeps small ponds wet in the late spring and earliest summer. The easternmost spring is at the base of the basalt caprock, similar to the springs surrounding South Table Mountain.

What about the "pure Rocky Mountain spring water" used to brew beer at Coors Brewery (B on Fig. 63)? Several websites report underground springs as the source of brewing water. One claims that 70 to 80 springs exist in the Golden area. Another claims that Aldoph Coors found water bubbling up

Fig. 64. 1872 photograph of Golden shows a dark patch of trees (below arrow), probably fed by springs at the base of South Table Mountain. View to east. Wm. Henry Jackson (1872) USGS online archive.

from springs in the Clear Creek Valley.[158] Perhaps 70 to 80 springs is an exaggeration, but apparently the Coors brewhouse is situated on top of historic springs located at the base of South Table Mountain. An 1872 photograph of Golden (Fig. 64), pre-dating the brewery and the Welch Ditch, shows a large patch of trees at what became the brewery area. Such a large tree-patch in an arid climate suggests a significant, constant water-source such as a spring. Were there truly springs in Clear Creek when Adolph Coors arrived in 1873? Yes, and the water levels in those springs declined as withdrawals for brewing increased over the years. Today, groundwater seepage and/or diversions from Clear Creek fill former sand and gravel pits east of the brewery that now are used for brewing purposes.[159]

---

[154] State of Colorado, 1882, p. 11. Local resident and environmental engineer, Frank Blaha reports old pipes at that location, too.
[155] CT 30 April 1931.
[156] State of Colorado, 1906, p. 3.
[157] Golden Landmarks Association http://goldenlandmarks.com/history-of-the-pullman-house/
[158] See https://thehistorybandits.com/2016/03/18/the-coors-banquet-beer-and-its-nostalgic-allure-a-case-study-in-history-marketing/ and https://biography.yourdictionary.com/adolph-coors
[159] Arbrogast et al., 2002, p. 27.

Chapter 8: Going for the Gold

Golden has a long, rich mining history that continues today. Individuals seeking their fortunes during the 1859 Colorado (aka Pikes Peak) Gold Rush founded Golden. Perfectly situated along a creek at the mouth of the canyon leading to the gold fields to the west, Golden began life as a mining camp. It quickly became a trade center linking the gold fields to the settlements of the Great Plains, including, of course, Denver. Golden grew with fits and starts due to the Civil War and the Financial Panic of 1873. But full of natural resources, literally within the town limits, Golden became a mining center for coal, clay, stone, and limestone. Five smelters in Golden processed ore from the gold fields that, along with other industries such as brick-making and beer-brewing, required coal for fuel. In the late 1800s, Golden was a small mining and industrial town with mining-related businesses, most of which have vanished. That story unfolds in this and the following chapters.

Was Golden named for gold? No. At first called Golden City, entrepreneurs of the Boston Company founded Golden on on 16 June 1859. Company members included George West, later publisher of the Colorado Transcript, James McIntyre, and F. W. Beebee, who began surveying the town site, a task later completed by civil engineer, Edward L. Berthoud. Likely Golden City got its name after businessman Thomas L. Golden.[160] But, there was and still is, a little bit of metallic gold in Golden.

Gold in Golden?

Gold in Golden? Yes, a little. Gold flakes, known as placer gold, are present in the Quaternary alluvium in the bottom and along the sides of Clear Creek and perched along the hills above Clear Creek. Gold flakes lie among the cobbles of the alluvium. All alluvium washed down Clear Creek by glacial flooding (Chapter 6) has gold flakes in it, because it was derived from gold-bearing bedrock upstream. Early miners exploited that fact. The mining process to extract alluvial gold is known as placer mining, which in its simplest form is gold panning. Notably, all placer mining requires a nearby source of water to conduct the mining operation.

Fig. 65. Hard work gold-panning on Clear Creek! X-60186 DPL WHC, used with permission.

Gold panning by hand (Fig. 65) is a slow way to recover gold from alluvium. Faster ways involved hydraulic mining and dredging. In hydraulic mining, water under pressure is jetted into the gold-bearing sand and gravel which then is eroded away through a sluicebox. The heavier gold is collected from the bottom of the sluicebox, and the sand and gravel ends up somewhere downstream. Dredging consists of digging up the gold-bearing sand and gravel with machinery that runs the alluvium through a sluice, separating the gold and returning the waste-rock to the stream banks, leaving behind large piles of sand, cobbles, and gravel. Both methods churn up the landscape and leave what we would now recognize as an ecological nightmare.

For short times between 1885 and 1892,[161] hydraulic mining was conducted along Clear Creek at two areas around Golden (Fig. 66). One operation was at the mouth of the canyon on the south side of Clear Creek, just below Welch Ditch. F. D. Benjamin, an

---

[160] Thomas Golden and George Jackson had camped for the winter in 1858-59 on Clear Creek in the area that is now Golden. In early 1859, Jackson and Golden proceeded up Clear Creek. Tom Golden became more interested in hunting elk, and Jackson continued his search for gold, which he found near today's Idaho Springs (Simmons, 2004 and Kile, 2002). Upon Jackson's return to the base camp, Jackson and Golden had a falling out. Thomas Golden later set up a storage and transfer business in Golden Gate City along today's Golden Gate Canyon Road, a mile north of Clear Creek. (Golden History Museum Epic Events, 2021).
[161] Gardner, 2004, p. 27.

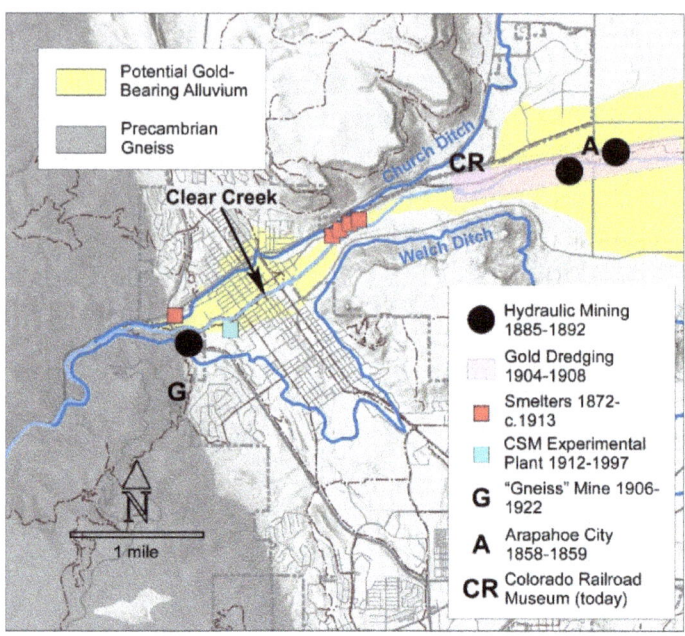

Fig. 66. Placer-gold mining, smelters, historic places, and CSM mining facilities in the Golden area, 1858-1997.

employee of Charles Welch, used water from Welch Ditch. With a vertical fall of 75+ feet to generate water pressure, Benjamin ran water through a canvas hose down to the creek, jetting out the gold-bearing Quaternary alluvium in the bank along Clear Creek.[162] The watery mixture was run through a sluicebox, and gold was collected at the bottom, as water, sand, and gravel bypassed into the creek. The operation did not last, though, due to the small size and quantity of gold flakes, as well as competing with water needed for irrigation. But it likely made an unsightly mess on Clear Creek through Golden for a while.

The other hydraulic mining area was 3 miles east of Golden at Arapahoe Bar,[163] an original site for finding placer gold along Clear Creek in 1858. Hydraulic mining took place at two sites. One site was on the lands of J.E. Wannemaker, who sold his land to a company with experience in hydraulic mining from the gold fields of California. This operation was a bit more complicated than simple sluicing. With a fall of about 175 feet to generate pressure, water from Church Ditch was piped to a hydraulic nozzle that tore up the banks and washed the gravel and gold to the foot of a hydraulic elevator. The water-gravel mixture under pressure in the elevator was then carried upward where it became sorted by gravity. The largest and heaviest boulders fell down first and were removed to a nearby dumping area.[164] The remaining mixture ran through a sluicebox to concentrate and gather gold flakes. Although the initial excitement was palpable, the operation proved uneconomic due to finding very small gold flakes in low quantities, along with a lot of huge boulders. About a mile upstream from Arapahoe Bar, on a site called Lewis Bar, another hydraulic mining elevator-operation used water from Welch Ditch with a fall of over 200 feet. The operations were abandoned after 1886.[165]

Fig. 67. Eleanor #1 dredge with North Table Mountain in background. View to north near today's McIntyre Street. X-60126 DPL WHC, used with permission.

The lure of placer gold in Clear Creek alluvium did not diminish over time. Two large gold dredges, the Eleanor #1 (Fig. 67) and #2, operated from 1904 to 1908 along the banks and in the channel of Clear Creek beginning near today's I-70 and moving upstream to beyond McIntyre Street, west of today's site of the Colorado Railroad Museum (Fig. 66). Each machine was capable of moving 3000 cubic

---

[162] CT 22 July 1885, 2 September 1885, 3 March 1886, Colorado Mining Gazette 19 May 1886. The gold-bearing gravel bars along Clear Creek were also called Lewis Bar and Wannemaker's Bar. The "town" of Arapahoe City existed near the diggings from 1858 to 1859. Golden was a much better location for a settlement. A historic marker for Arapahoe City exists west of McIntyre Road.
[163] Parker 1974, p. 75 and CT 21 October 1885.
[164] CT 21 October 1885.
[165] Curiously, federal law banned hydraulic mining in California in 1884 because of devastation of farmland and fish habitat caused by sediment in streams and rivers. Maybe that is why so much equipment was available to use in Colorado, after 1884! An 1893 federal law re-established hydraulic mining in California if the operator constructed sediment-retention structures.

yards of alluvium daily.[166] As with hydraulic mining, dredging the Clear Creek alluvium became unprofitable due to the small flakes of gold relative to the large number of cobbles and boulders that had to be discarded. Dredging also was stopped when large basalt boulders shed from the Table Mountains became abundant.[167] The Eleanor #1 was dismantled and sent to French Gulch near Breckenridge, where it created the cobblestone "dunes" composed of tailings, still plainly visible. It remains in Breckenridge where it forms the ruins of the Reiling Dredge. The Eleanor #2 was sent to Sacramento, California. A few gold-dredge cobble piles are still visible on the north side of Clear Creek along the regional trail west of the intersection of I-70 and 44th Street in Wheat Ridge.

In 1922, a quarry operator in Adams County placed a "gold-saving" device (probably a sluicebox) on a sand and gravel washing operation along Clear Creek east of the Coors Brewery.[168] From then through the 1970s, placer gold recovery enhanced the economics of sand and gravel quarrying along Clear Creek east of Golden. Small-scale gold panning was rejuvenated during the Great Depression, especially in 1933 after the U.S. curency went off the gold standard, and the price of gold increased.[169] Not easy money, but during those tough economic times, gold panning was a way to make a few bucks. Gold panning stopped during World War II when it was banned by federal order. After the War, panning became more of a hobby-oriented operation.

Upstream from Golden, in Clear Creek Canyon Open Space Park, a ten-mile stretch of the Creek is open to gold prospecting without a permit and by adhering to simple rules. Within Golden, prospectors still pan for gold along the Clear Creek Kayak Run during low-flow periods. Another historical placer mining area open to gold panning is at Arapahoe Bar along Clear Creek southwest of the intersection of I-70 and 44th Street in a City of Wheat Ridge park.

Golden's One Gneiss Mine

Golden had one hard rock mine, of sorts. In 1906, an adit a couple of hundred feet long with a couple of side tunnels was dug straight into the Precambrian gneiss in the mountainside below Mount Zion (Figs. 66 and 68). The non-existent "ores" of Precambrian gneiss (Chapter 2) were worthless. But at the time, a mine construction project near CSM was considered a priceless educational opportunity.

The "gneiss" mine was built to conduct "critical and comparative studies of process and machines" and to "solve problems in mining engineering."[170] The mine location is north of the Lower Chimney Gulch trail. The old adit and side tunnels have been part of a private residence since

Fig. 68. Colorado School of Mines "tunnel" house at the "gneiss-mine" opening (arrow), circa 1908, with Mt. Zion on the skyline. View to west. Arthur Lakes Library, Colorado Digitization Project, Russell L., and Lyn Wood Mining History Archive, used with permission.

---

[166] Gardner, 2004.
[167] Parker, 1974, p. 123. The presence of basalt boulders, derived from North Table Mountain, indicated a loss of gold-bearing granitic rock.
[168] Ibid. p. 22.
[169] Ibid. p. 20.
[170] Borowsky, 2015.

the 1950s. The Edgar Experimental Mine in Idaho Springs[171] that CSM acquired in 1921 for educational purposes superceded its use.

Smelters in Golden

"There's gold in them thar hills." In 1859, gold mining in the Idaho Springs-Black Hawk-Central City area began as placer mining in alluvium along the upper reaches of Clear Creek. Miners soon traced the gold flakes to their bedrock origin, thus leading to "lode" mining by tunneling into mountain sides. But, finding gold veins ("the Mother Lode") was relatively easy compared to the difficulty of extracting the gold encased in sulfide ores. In 1867, Nathaniel Hill, using a Welsh smelting process, applied his research to his Argo Smelter in Black Hawk.[172] Successful smelting brought new life to Colorado mining. It also incentivized William Loveland to construct a narrow-gauge railroad, part of the Colorado Central Railroad (CCRR), to transport ores[173] to Golden, operating by 1872. When it came to getting milled ore to Golden, Loveland had a one-track mind.

Fig. 69. The Golden Smelter on the north side of Clear Creek (foreground) with a train on the Colorado Central RR tracks, circa 1875. North Table Mountain in background. View looking north. Golden Landmarks Association, used with permission.

Five smelters sprang up in Golden (Fig. 66), with ore-a-plenty coming out of the mountains and the CCRR connection to Denver established in 1870. The first smelter in Golden, the Golden Smelter (1872[174]), smelted gold, silver and copper (Fig. 69). The Malachite Smelter (1878, aka Valley Smelting) focused on copper. The Collom Company (1875) smelted low-value lead, while the French Smelting Works (1877) and the Trenton Dressing and Smelting Company (1878) smelted a variety of ores. Four of the smelters were on the east side of Golden along "Smelter Row," stretching out along the north bank of Clear Creek northeast of the Coors Brewery, one right next to the other. The Trenton Dressing and Smelting Company was on the west side of Golden on the southwest side of 8th Avenue just east of where 8th Ave. intersects the Church Ditch. All the smelters were next to the CCRR tracks.

Silver bullion was by far the biggest smelting output by dollar-value, with lead associated with silver in the ore, being the largest output by volume. In addition to its railroad connections, Golden was an obvious choice for smelting as it produced readily available brick for the smelter buildings, heat resistant fire-brick for smelter furnaces, and coal for fuel in the smelting furnaces.

For a short time, Golden could legitimately claim that more ore was smelted in Golden than in all the rest of Colorado. As an added benefit, the engine room of the Valley Smelter served for a time as Golden Town Hall; the storage area for Fire Department apparatus; the street commissioner's wheelbarrows; and other town needs. It exemplified the motto on Golden's coat of arms at that time: "Nothing if not useful."

---

[171] The Edgar Mine in Idaho Springs is part of the CSM campus, and for a nominal fee, public tours are available by reservation: call +1-303-567-2911.
[172] Nathaniel P. Hill (1832-1900) resigned his Brown University chemistry professorship and came to Colorado in 1865 to try his hand at mining. Seeing the need for better smelting techniques to handle the sulfide-rich ores, Hill adopted a Welsh smelting technique and opened the Argo Smelter in Black Hawk in 1867. Hill was a US Senator from Colorado (1879-1885) and ended his career as a Professor of Metallurgy at CSM, where Hill Hall bears his name.
[173] The raw ore mostly was crushed, or milled, near the mines in the Idaho Springs, Central City, Black Hawk area, and the milled ore was transported to Golden.
[174] Although plans for the smelter began in 1868, the first bullion was produced in 1872. CT 4 September 1872.

However, as silver was removed as a monetary standard in the United States in 1873,[175] the value of silver dropped, and mining slowed. Another blow was the Sherman Silver Purchase Act of 1890, which had guaranteed a high level of silver purchase by the U.S. Government. Instead of stimulating the market, the resulting over-production of silver further lessened its price, leading to the Silver Panic of 1893. Repeal of the Act in 1893 removed price supports and further killed silver mining in Colorado. One by one, the smelters closed. Only the Golden Smelter re-opened from 1910 through 1913 until it went into receivership and closed for good.[176]

CSM Experimental Plant (CSMRI)

In 1912, the Colorado School of Mines built a commercial-sized experimental ore processing and metallurgical plant, dubbed the "Experimental Plant," (Figs. 66 and 70) on the south bank of Clear Creek in the area of today's CSM soccer field. At the time, low-grade ore, left after mining the "good stuff," required a lot of processing to exploit economically. In an early version of public-private partnerships, V.C. Alderson, then-president of CSM, saw a great need for a facility that could "call the attention of mining men to the necessity of a careful preliminary investigation of their ore before they erect a costly mill."[177] The state-funded facility was a separate department at the school where students and faculty conducted ore-processing experiments with supplementary funding from private companies. Throughout the 1920s and 1930s, experiments included processing low-grade lead-zinc-silver ores, radium, and molybdenum; crushing coal and rock; retorting oil shale; and refining petroleum. Facilities expanded in the 1930s to include a U.S. Bureau of Mines facility to conduct processing experiments with coal from all over the U.S.

Fig. 70. Experimental Plant (CSMRI) circa 1993: aerial view looking east over the red-roofed buildings of the Plant. Clear Creek channel in middle ground. Sanborn Company, used with permission.

In 1949, plant operation was transferred to a separate organization called the Colorado School of Mines Research Foundation, funded solely by private industry.[178] The facility was also renamed the CSM Research Institute (CSMRI). Post-war experiments included uranium ore processing; oil-shale processing; cyanide leaching and other metals-related processing experiments. In 1987, the Research Foundation was disbanded, and the facilities closed. The idled buildings were removed in 1997. The contaminated physical site was remediated three times between 1987 and 2016 (Chapter 7).

---

[175] The Coinage Act introduced by U.S. Sen. Sherman in 1873 ended silver as a "bimetal" monetary standard in America, putting the nation entirely on the gold standard (Silber, 2019).

[176] The Guggenheim Family patriarch, Meyer Guggenheim (1828-1905), owned several smelters in the U.S. including the Omaha and Grant Smelter in Denver. Buying the last smelter in Golden, Guggenheim shut it down as the cost to operate it was greater than their Denver smelter. Notably, Meyer's son, John Simon Guggenheim (1867-1941), lived in Denver and celebrated the birth of his first son, John, in 1905 by donating $80,000 ($2.2M in 2019 dollars) to CSM for construction of its iconic gold-topped Simon Guggenheim Hall. Simon was elected as a US Senator from Colorado 1907-1913.

[177] Alderson, 1912, p. 269.

[178] Oredigger 13 February 1951.

Chapter 9: Black Diamonds

Coal lacks the romance of gold and other metals. But good grades of coal have a shiny, black luster, leading to coal being called "black diamonds." The diamond-like luster indicates a high purity that leaves behind nothing but a white ash after burning: a very desirable quality! Advertisers selling Golden coal commonly used the phrases "white ash" and "black diamonds" to promote their product. Similarly, those names applied to coal mines helped advertise their product. Coal mining in Golden was most active between the late 1860s and 1900, helping to fuel Golden as an industrial center.

Coal Below the Ground

In the 1800s, prolific coal mines of the U.S. and Europe were sited in flat (un-tilted) coal seams. Digging a mine in tilted beds, or seams, like those along the Golden Fault System (Chapter 5), initially was feared to be a bad idea. As stated in 1870,

> when the railroad is finished to Central [City], we may possibly be induced to visit Golden City. We will certainly come if Geo[rge]. West will take us to the coal mine and show us the grand spectacular drama, of raising 1000 tons of coal a day from a perpendicular coal mine. The fact is, Golden City coal beds…have the very uncomely fashion for coal mines of standing up on end.[179]

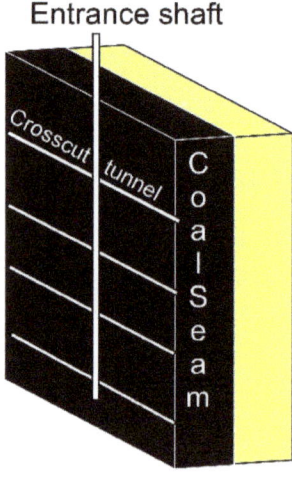

Fig. 71. Schematic showing how vertical coal seams were-mined. See text for discussion.

That skeptical comment ignored the fact that metallic ore deposits with similar geology, such as vertical veins, had been worked for centuries throughout the world. With a big demand for local, good-quality coal in Golden, the technical problem with "<u>uncomely</u> vertical coal mines" was quickly surmounted, especially with the help of emigrant miners like John Nicholls[180] who had previous metals and coal-mining experience in England and the eastern U.S. The invention of powerful steam-lifted hoists solved the problem of lifting heavy loads.

Underground coal mines in Golden's vertically tilted seams used a main vertical shaft, either sunk directly on the outcropping seam (Fig. 71) or a short distance away, in which the main shaft then connected to the coal seam by one or more short tunnels. Miners dug more tunnels, called crosscuts, by pick and shovel along (parallel) to the seam and propped up with logs to help prevent cave-ins. Coal could be extracted laterally along the crosscuts for as far as the mining rights extended to property lines or as far as a coal seam extended laterally. To extract the most coal, miners dug successively deeper levels, leaving a vertical distance, usually at least 100 feet, between the crosscuts to avoid collapse of the overlying coal. For safety reasons, then, the vertical space between crosscut levels left-behind a lot of coal. Most Golden coal mines extended less than 300 feet deep, except for one, the White Ash Mine, that was 730 feet deep.

Sound dangerous? It was. Not only were cave-ins a problem, the coal could catch on fire. Coal also contained associated poisonous gasses, such as methane, which could explode, and carbon dioxide, which suffocated the miners. Ever hear of the canary in the coal mine? When the canary could not breathe, you needed to get out…fast. Let's just say that 1880s safety standards were not the same as those now.

---

[179] CT 27 April 1870.
[180] Nicholls worked in metals mines as a youth in Wales. From 1863-1873 he worked in the coal fields of Pennsylvania. Moving to Golden, he established several Golden coal mines (Baskin and Millet, 1880, p. 582-583).

Golden Coal Mines

With the presence of coal seams in the Laramie Formation (Chapter 4) and the demand for local fuel for homes, steam locomotives, smelters, and brick kilns, coal mining in Golden was a natural pursuit. A thick, semi-continuous coal seam trending south to north on the west side of town led to several coal mines that supplied coal within Golden, to Denver, and to the mountain mining towns via the CCRR. Because of historic demand for Golden's good quality coal, more than 13 coal mines operated at various times across Golden from the late 1860s to 1940, all located along the Laramie Formation outcrop belt (Fig. 72).

The first coal mines in the greater-Golden area opened in the late 1860s. These included the Murphy and Tindall mines along Ralston Creek and the Ralston Springs/Ideal mines along Van Bibber Creek north of Golden. South of Golden, several coal mines operated on the western slope of Green Mountain, including the Wheeler and Rowe coal mines.

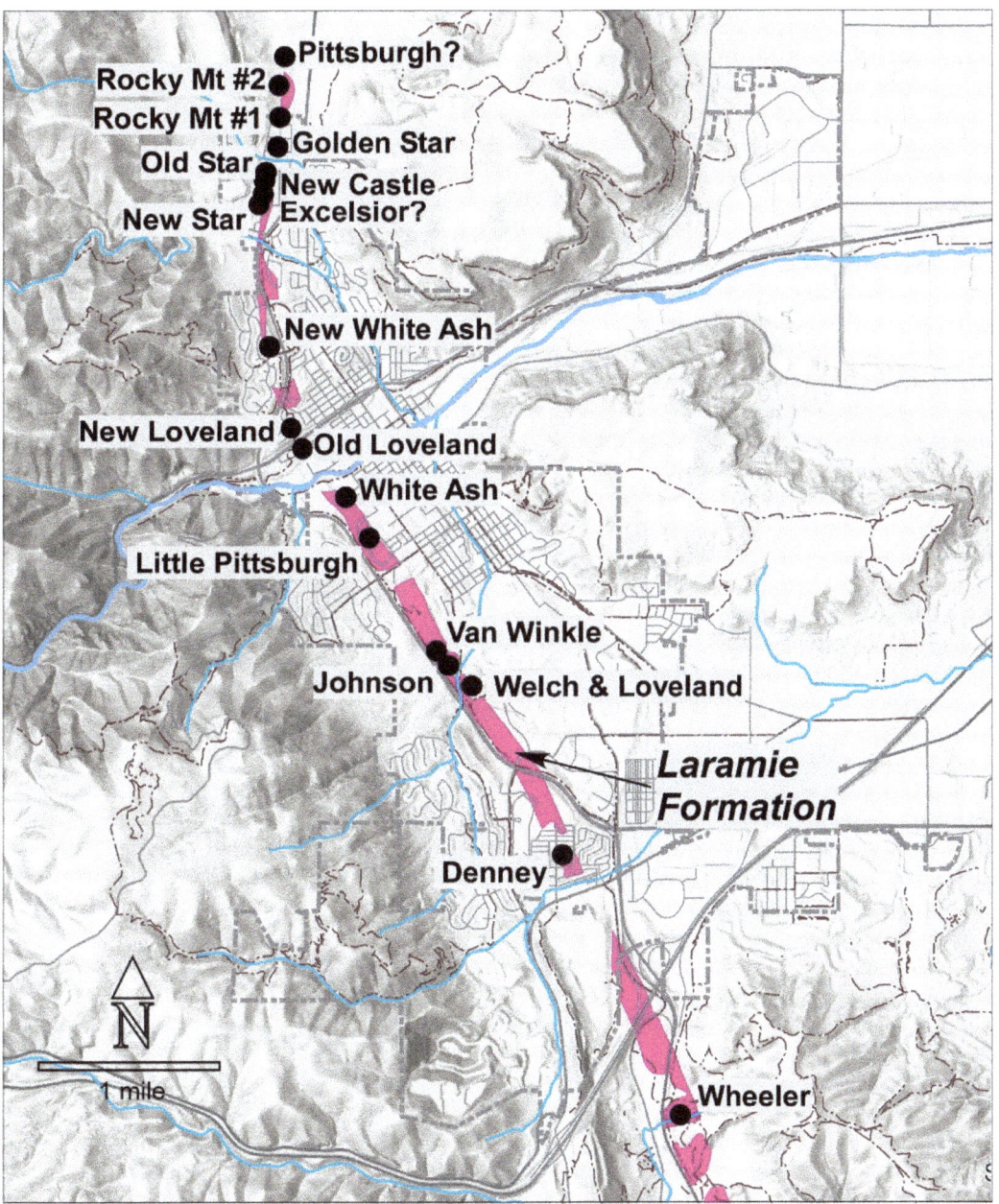

Fig. 72. Historic coal mines in the Golden area occurred in the Laramie outcrop belt (magenta). Mine locations after Amuedo et al. (1978), Carroll and Bauer (2002), Willits (1878 and 1899), and old topographic maps.

The main coal seam in Golden, known as the White Ash seam (Chapter 4), was discovered by Edward L. Berthoud and J.M. Johnson in 1862[181] within Laramie Formation outcrops on the west side of today's CSM campus (Fig. 72). From the surface downward for about 100 feet the coal was fractured and dull, and generally of poor quality, leading to some early arguments about the quality of Golden coal.[182] However, below about 100 feet deep, the coal was excellent. Black, lustrous, and hard, it looked like "black diamonds." It burned well with no bits of shale ("clinkers") resulting in a "white ash," yielding the name of the White Ash coal-seam and the names of two Golden coal mines.

An early coal mine in south Golden, the Welch-and-Loveland Mine, probably opened around 1867 but was abandoned by 1870 after it was the site of a dumped body from a murder.[183] It was located in the area of the historic Hoyt's Ranch, at the southern end of today's Illinois Street (Fig. 72). The Johnson Mine (also known as the Prout Mine) was opened in 1874[184] in the same area, possibly supplying coal to the Cambria Lime Kiln after 1879. In 1874 or 1876, Evan Jones, John Hodges, and John Nicholls opened the Little Pittsburgh Mine located on the west side of the CSM campus at today's West Campus Rd.[185] Workings were about 100 feet deep, and like many of the coal mines in the Laramie Formation, later clay mining quarried through the old coal mines, destroying the old coal workings.

The Old Loveland Mine (Figs. 72 and 73; originally the Black Diamond Mine) was another early coal mine within Golden, possibly opened in 1867, with a shaft sunk to 125 feet by 1870.[186] Located on the White Ash coal seam, it was on the north side of Clear Creek just north of the current location of 1410 8th Street, now in the right-of-way of State Hwy-58. Started by several businessmen, John Nicholls took it over in 1878 until it closed in 1879.[187] The workings ultimately consisted of two levels down to 300 feet.

Fig. 73. The Old Loveland and White Ash Mines in 1890. View east. X-9799 DPL WHC, used with permission.

---

[181] Berthoud, 1880, p. 367.
[182] Ibid. Professor Mallet, the first "Professor-in-Charge" at CSM, and CSM students investigated the composition of the local coal and attested to its high quality.
[183] CT 19 October 1870.
[184] Golden Globe 14 February 1874.
[185] Slader, 1952, but Baskin and Millet, 1880, put the opening of the Little Pittsburgh Mine as 1876.
[186] CT 21 December 1870.
[187] Golden Globe 1 May 1893.

The original White Ash Mine, at the end of today's 12th Street (Fig. 73), was opened in 1874 by John Hodges, with initial workings up to 300 feet deep. It was operated by R.D. Hall and A.L. Jones beginning in 1877 until 1879 when Mr. Hall retired and W.S. Wells purchased an interest. Mine workings were deepened in 1885, when the Golden Fuel Company rejuvenated operations by entering into a 99-year lease with John Hodges and Charles Welch. By 1889, the workings extended to 730 feet, and the White Ash Mine became the deepest coal mine in Colorado. The White Ash and Old Loveland Mines had a checkered past, recounted below.

In north Golden, a series of closely-spaced mines followed the underground trend of the White Ash coal seam. Opened at various times beginning in the early 1880s, these included the New Star, Excelsior, Old Star, Golden Star, Rocky Mountain, and Golden Mines (Fig. 72). Mine workings ranged from 100 to 200 feet deep. All were north of Golden Gate Canyon Road, along a line beginning at today's New Star Way near today's City of Golden concrete recycling pile, then north to today's city limits. After abandonment, the ground above the mine workings experienced subsidence (Chapter 12).

Two former coal mines are now City of Golden parks. New Loveland Mine Park at 1301 Rubey Drive commemorates the New Loveland Mine (Fig. 74), and the New White Ash Mine Park at the northeast intersection of State Hwy-93 and Iowa Street commemorates the New White Ash Mine. Both mines opened in 1890 after the White Ash Mine disaster. They both closed by 1900. Almost 100 years later, the abandoned mine sites became parks in the early 1990s when former Coors property was developed into housing in the Canyon Point area. The parks provide a buffer to mine subsidence issues involving the abandoned underground workings.

Fig. 74. New Loveland coal mine circa 1890. Dark patches in left foreground are the workings of the Old Loveland (aka Black Diamond) Mine. View to northwest. Golden History Museum and Park, City of Golden Collection, used with permission.

After 1900, coal mining in Golden almost ended due to the expense of underground operations and competition from more prolific coal mining areas in Colorado. However, the first half of the 20th century saw the brief opening of two new mines. In 1903, the Van Winkle Mine opened near the former Johnson Mine along Kinney Run Gulch (Fig. 72), initially to sell coal to CSM. Workings went down to about 145 feet, and operations continued sporadically until the Great Depression. One completely new mine, the Denney Mine (Fig. 72), opened during the Great Depression in 1933. Located northeast of today's US Hwy-40 (Colfax) and Zeta Street, it created great hope, as it was touted to provide jobs when the Colorado unemployment rate was about 25%. But, it only lasted a year.

By 1940, no coal mines operated in Golden. Coal mining in the greater Golden area ceased entirely by 1950, with an estimated ten million tons of coal having been mined in the entire area.[188] With an average thickness of 6 feet of minable coal, 250 million tons of coal still lie within 1000 feet of the surface.[189] Recovering it now is unfeasible due to poor economics, urbanization, and changed attitudes towards coal as a sustainable fuel. Nonetheless, coal mining is a clear part of Golden's history, and the abandoned underground workings left a legacy in terms of mine subsidence (Chapter 12).

---

[188] Van Horn, 1976: the Leyden Mine, the last coal mine in the area, north of Ralston Creek, closed in 1950.
[189] Ibid. p. 103.

1889 White Ash Mine Disaster

The good burning properties of Golden's coal are reflected in the name of the most memorable, and perhaps infamous, Golden coal mine, the original White Ash Mine. The White Ash Mine portal was on the south side of Clear Creek at the west end of today's 12th Street. There, in the subsurface, below the alluvium of Clear Creek (Chapter 4), the coal seam tilted steeply to the east near the surface, then became vertical for more than 600 feet downward.

The White Ash Mine was plagued with problems, most of which were typical of the time. Mining paused in 1879 when a fire broke out at the 280-foot level (Fig. 75). That tunnel was sealed up, with the thought that a lack of oxygen would smother the fire. From 1885 to 1889 the mine had several accidents

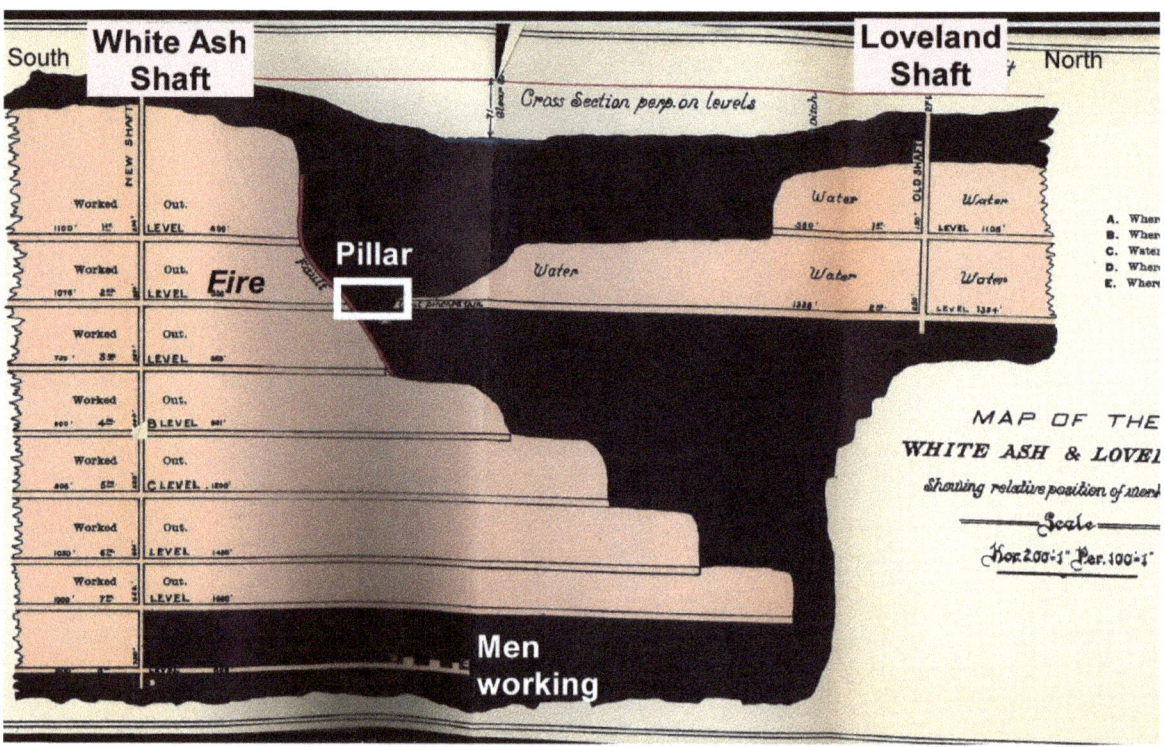

Fig. 75. North-south cross section (cut-away view) of the White Ash and Old Loveland coal mines showing configuration of underground workings in 1889. Diagram modified from McNeil (1890).

from falling rock and equipment, as well as bad air from carbon dioxide gas and fires within the mine tailings at the surface (e.g., Fig. 73, to left of White Ash Mine buildings). As the workings reached the 720-ft level by 1888, Colorado State Mine Inspector and engineer, John McNeil visited the mine several times. Finally, in 1889, he required replacement of unstable timbers, as well as the drilling of a 700+ foot deep escape shaft at the north end of the mine, north of Clear Creek. He also restricted the number of miners to ten for any given shift for safety reasons.[190] The mine owners replaced the rotten timbers and began to arrange for a second shaft in the summer of 1889. In a letter dated July 12, 1889, twelve miners petitioned the State of Colorado to keep the mine open due to hardship,[191] and the mine continued operating.

Meanwhile, the Old Loveland Mine, (aka Black Diamond Mine) was about 1960 feet north of the White Ash Mine (Fig. 75), where the White Ash coal seam extended under and across Clear Creek. The Old Loveland Mine had been abandoned in 1879 due to suffocating inert gases, carbon dioxide, aka "black damp," at the 250-foot level. The abandoned Loveland shaft and tunnels were left to fill by water seepage. In 1889, the flooded mine was being used as a water source for the steam boilers at the nearby

---

[190] McNeil, 1890, p. 9-16.
[191] Ibid. p. 15.

Golden Brick Works. A pillar of rock and coal 70-100 feet thick separating the abandoned tunnels of the Loveland and White Ash Mines (Fig. 75) had been considered "safe enough" by Inspector McNeil and others.

On 9 September 1889, at about 4:00 p.m., after the evening shift had gone down to work in the active level [720 feet], Charles Hoagland, mine engineer, reported:

> *The men below sent up a signal to send down the cage. It was at once lowered. It went to within about 6 feet of the bottom of the [vertical] shaft and the struck something...We worked with the cage for a few minutes, and finally something below broke. Since then we have heard nothing...[The foreman, Evan Jones] went down on a ladder to the 280-foot level and came back...the air was so bad he had to ascend...[The men] went over to the old [Loveland] mine at 6 [p.m.] and found that shaft to be perfectly dry. [It was always filled with water, used at the Golden Brick Company].*[192]

People gathered at the White Ash Mine portal throughout the evening and night, crazy with grief. Inspector McNeil was called at 9 p.m. When McNeil arrived at midnight, the shaft was full of carbon dioxide, and the water-level in the main shaft was 100 feet above the bottom, at the 680-foot level. After ventilating the shaft by 8:00 a.m., McNeil and mine-foreman Evan Jones went down the shaft in a bucket. On the way down they encountered intense heat and could see timbers on fire at the 280-foot level. At the 480-foot level,[193] they could hear water running. The mine workings were filling with water that could not be pumped out.

Inspector McNeil formally determined that the rock and coal pillar separating the Old Loveland and White Ash mines broke, likely weakened by a coal fire, allowing water to burst through the blocked-off 280-foot tunnel of the White Ash Mine. Water flowed quickly across the tunnel to reach the vertical shaft and then down to the 720-foot level, filling the side tunnel and drowning the ten miners. The event was the biggest disaster in Golden's mining history.

Thereafter, the White Ash Mine was abandoned and at one point in time, was considered a cemetery.[194] Until the 1950s, on the anniversary of the disaster, Goldenites placed flowers and held a memorial at the former mine site. On 9 September 1936, in commemoration of the lives lost in the tragedy, a monument was placed at the west end of the CSM football field near the mine shaft. The

Fig. 76. Statue near west end of 12th Street honors the ten miners who drowned in the White Ash mine disaster on 9 September 1889. View south.

ceremony, attended by 1000 Goldenites, was led by Mayor Bert Jones, the son of former mine foreman, Evan Jones, who sought to rescue the trapped miners 47 years earlier.[195] A new memorial (Fig. 76) was dedicated by former Golden Mayor Marv Kay on 29 October 2016. It is on the CSM campus across from the football stadium (Marv Kay Stadium) near the west end of 12th Street, where it is Stop 7a on the Weimer Geology Trail (Chapter 4).

---

[192] McNeil, 1890. The WPA history of Golden (Foothills Genealogical Association of Colorado, 1993, p. 474-476) has the full account from the Golden Globe. Issues recounting the disaster in the Colorado Transcript and Golden Globe are lacking in the collections of Colorado Online Historical Newspapers.
[193] McNeil and Jones showed incredible bravery!
[194] Shown as a cemetery on the USGS Golden Quadrangle topographic map of 1965.
[195] CT 10 September 1936.

Chapter 10: From a Lump of Clay

For over 100 years, Golden was a major center of clay mining in Jefferson County and Colorado. According to the U.S. Geological Survey, more than half the clay mined in Colorado came from Jefferson County, most of it from Golden, amounting to many millions of tons of clay. Clay, in reality, established the only Golden-based industry still remaining from its mining heritage, putting Golden on the map, so to speak.

Clay's Anatomy

What is clay? Clay refers to both a particle size and a mineral.[196] As a particle size, clay particles are less than 8 microns in diameter (very tiny). Clay minerals have a distinctive sheet-like molecular structure, or crystal lattice. All clay minerals are aluminum silicates: aluminum (Al) combined with silicon (Si) and oxygen (O). As it turns out, clay particles and minerals occur in sedimentary rocks called siltstone and shale, both of which are common to several rock formations in Golden.

Surprised to find aluminum in clay? The more aluminum in clay, the better for using it for pottery and high-tech ceramics. The most aluminum-rich clays, such as kaolinite, contain only aluminum, silicate and four hydroxyl (OH) groups. More impure clays have extra elements in the crystal lattice, such as iron, magnesium, potassium, calcium, and sodium, and lots of hydroxyl groups plus water (H_2O). The large spectrum of clay impurities makes the value of any given clay deposit range from excellent to poor. Golden's clay deposits were considered very good to excellent. They contained refractory and non-refractory clays, depending on the geologic formation where they occurred. The term "refractory" refers to the degree of heat resistance of a clay, which in turn governs the use of that given clay.

Refractory clays, or "fire-clays," withstand high firing temperatures above 1650°F.[197] Such high quality clays, mostly composed of aluminum-rich kaolinite, are used for fire-brick, pottery, such as pitchers, flower pots, chamber pots, and floor tiles, and since the early 1900s, high-tech ceramics. Fire-clays and their products were shipped by railroad to local and nationwide markets.

Clays with more impurities are "non-refractory," or brick-clays,[198] meaning that they fire or harden temperatures below 1650°F. The more "stuff" in the clay mineral like iron and magnesium, the poorer the refractory properties and the lower the value of the clay. Non-refractory clays are best for making-bricks and sewer pipes, hence the name "brick-clay." Due to the low value of the raw material and its products, bricks cannot justify excessive mining and transportation costs. Easy accessibility to local brick-clay deposits, coal to fire the kilns, and a rail-line to Denver prompted development of Golden brickworks.

Converting clay deposits to commercial use requires crushing, separating, and re-mixing a purified clay product. The mixed lump of clay is then shaped and fired in kilns. Because of the complicated commercial processes, recipes or formulas for clay products are closely-guarded trade secrets.

Clay Stoping and Open-Pit Mining

At first, mining Golden clay was mostly an underground operation. Having worked in the underground Murphy coal-mine along Ralston Creek north of Golden, George W. Parfet began the first large-scale clay mine in Golden in 1877. Naturally, he used similar underground mining methods for vertical clay beds as he had learned for vertical coal seams. However, the mining method for clay beds relied more on a process called "stoping." As a result, clay mines reached about 100 feet deep, much shallower than coal mines, as explained below.

---

[196] This is certainly confusing, and it confuses many geologists. But it is important to understanding clay's use. If you are making pottery, you really do care about this.
[197] Common in the era of outdoor plumbing, essentially before the 1930s.
[198] Remember the clay you used in grade school to make pottery? It was fire-clay that fired around 1900°F.

Underground clay mining in Golden involved first driving a horizontal or gently sloping entrance tunnel, called an adit, across the vertical sandstone beds to a vertical clay bed, then blasting away the roof of the clay bed up to the ground surface (Fig. 77). W.G. "Chip" Parfet, Jr. (great-grandson of G.W. Parfet) recalls the method used around 1900:

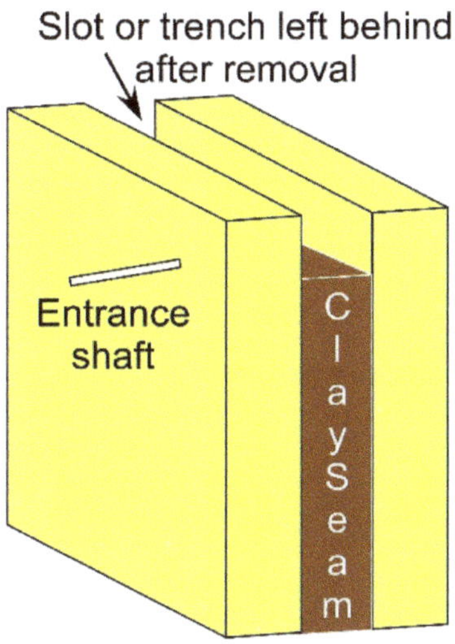

A tunnel was driven laterally to the clay seam: it was very hard drilling [through the vertical sandstone beds]. The standard was a tungsten-bit auger starting with 2 f[oo]t diameter then going to a larger diameter. The rule of thumb was to drill 8 holes, 8 feet deep in 8 hours to make the tunnel to the clay seam. Then the tunnel was shot [blasted] and cleaned out. Then at the clay seam they drove the rise [called a stope] upward into seam the same way, shot it, and cleaned out clay.[199]

Fig. 77. Schematic of a clay stope, which leaves behind a slot or trench at the ground surface after clay removal.

As the clay beds were mined out, the hard sandstone layers between clay beds became empty slots propped apart with logs (Fig. 78). The logs acted like audible "strain gauges." If the walls began to cave-in, the logs would start to bend, making a ringing sound that the miners could hear and know when to escape.[200] Clay was removed from the mines via ore carts and transported via wagons and/or railcars (Chapter 12) to one of several brickyards in Golden and Denver.

After World War II, surface, or open-pit, mining increased with the greater availability of mechanized equipment and as underground mining diminished due to high labor costs.[201] Dakota fire-clay was mined underground until about 1955. After World War II, with gasoline-powered shovels, brick-clay mining in the Laramie Formation turned exclusively to open-pit methods. Left behind after open-pit mining, the remaining sandstone beds between the clay seams form the "now-scenic" sandstone fins such as those on the Fossil Trace Golf Course and at the Jefferson County Laramie Building (Fig. 25 in Chapter 4).

Golden Clay Mines

Golden was a center of clay mining in northeastern Colorado for nearly 140 years. In 1900, Golden was touted as THE center of clay mining in Colorado.[202] Clay came from two geologic formations: the Dakota Group and the Laramie Formation. The clays in the two formations had different mineral compositions, used for making different products. Fifteen clay mines existed in and around Golden (Fig. 79), only one of which is now active.

Fig. 78. Log "strain gauges" at the Hoyt Mine in 1901 with a man for scale. Photo from Geijsbeek (1901, p. 243).

---

[199] Chip Parfet personal communication, 2 April 2020.
[200] Wall cave-ins were not that common, as many vertical walls are still standing today. Roof cave-ins, however, were a constant safety problem (see Chapter 12).
[201] Amuedo et al., 1978, p. IV-5, and Parfet and Ryder, 1977. Some surface mining occurred before World War II, but after that, surface mining was by far the most common mining method.
[202] Geijsbeek, 1901, p. 424.

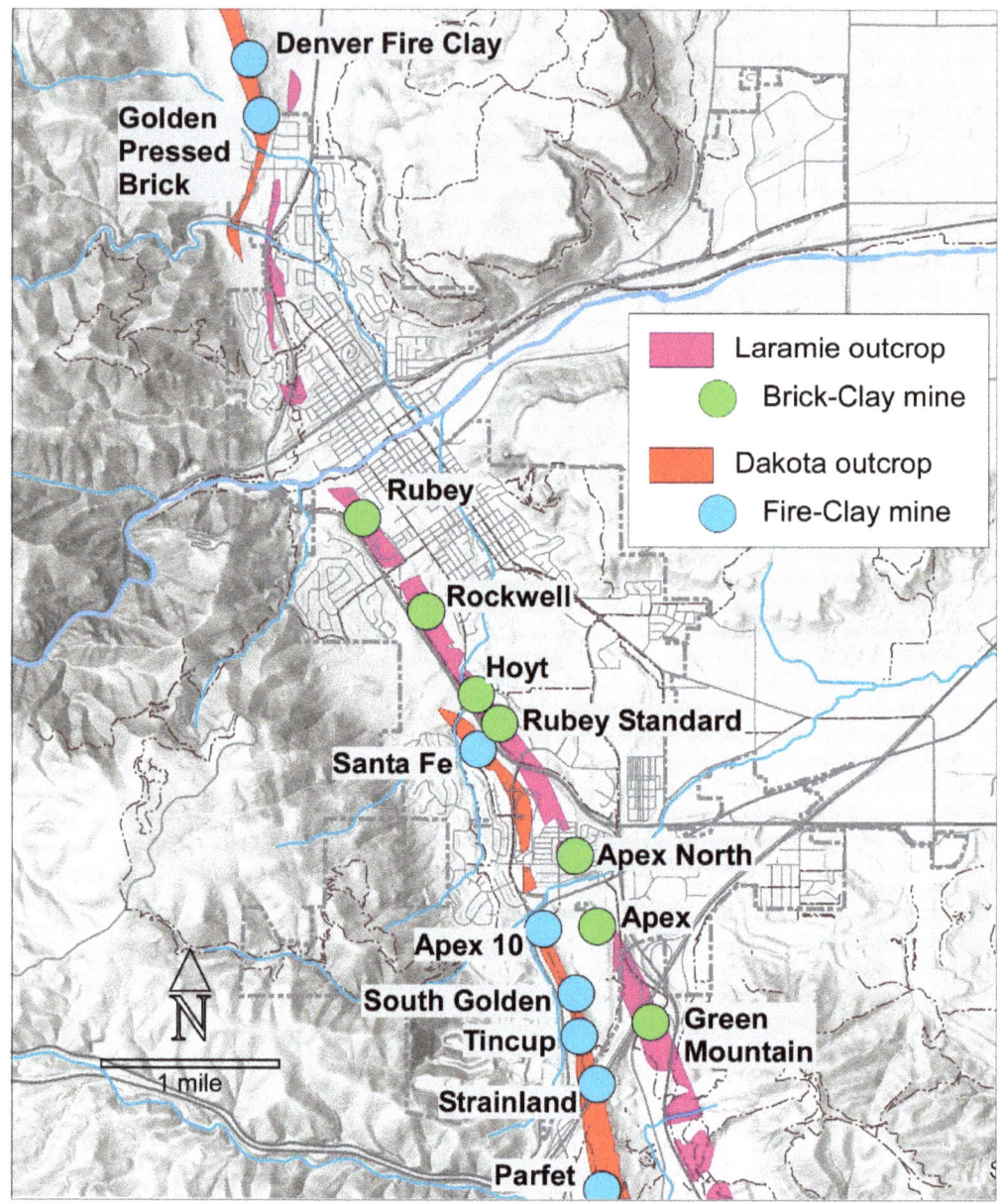

Fig. 79. Historic clay mines in the Golden area. Locations compiled from Geijsbeek (1901), Patton (1904), Van Sant (1959), Amuedo et al. (1978), old topographic maps, and aerial photographs.

Dakota Fire-Clay

The Dakota Group, specifically the South Platte Formation (Chapter 3), is rich in high-refractory kaolinite, the main component of fire-clay. Thus, the earliest fire-clay mining, around 1866,[203] in Golden likely was along the northern Dakota Hogback (Pine Ridge) in the vertically upturned outcrops. Today, trench "scars" along the the north and south Dakota Hogbacks (e.g., Fig. 80) are some of the most visible mining features Golden. Two long, white trench-scars from the Denver Fire-Clay Mine (now called the Golden Mine) are especially prominent in north Golden as one drives along State Hwy-93 (Fig. 80). The Golden Pressed Brick Mine was one of the earliest clay mines, followed by the Denver Fire-Clay Mine that eventually took over operations along the entire north Hogback. The Denver Fire-Clay Mine became

---

[203] Waage, 1961, reports that the date is difficult to pinpoint, but by December 1866, the Colorado Transcript advertised pottery made by Henry Bell. With no rail connections to Golden at that time, one can safely surmise a local fire-clay source that was mined from the nearby Dakota Hogback north of Golden.

the currently active Golden Mine in 1941 with a perpetual lease issued by the Colorado State Land Board. The mine lease is a special situation as it is within section 16, owned by the State of Colorado, with revenues for the benefit of schools. Today, it is operated by the Denver Brick Company, a subsidiary of Acme Brick Company.

Fig. 80. Clay-trench scars of the former Golden Pressed Brick and Denver Fire-Clay Mines remaining after stope mining along the Dakota Hogback (Pine Ridge) in north Golden. View to northwest.

In south Golden, the Santa Fe Mine (Fig. 79) has two moderately visible trench scars along Eagle Ridge,[204] northwest of the intersection of US Hwy-6 and Heritage Road. Operations at the Santa Fe Mine likely began in 1879, continuing until around 1905.[205] At the southernmost edge of Golden, former clay mining trenches are prominent along the east side of the Dakota Hogback (Tincup portion). They are especially visible travelling westbound on I-70, on the east side of Hogback at the I-70 roadcut. Several Dakota fire-clay mines existed in this south Golden area: Apex No. 10, South Golden, Tincup, Strainland, and Parfet Mines. The oldest mines are the Apex 10, South Golden, and Tincup Mines that began operations in the late 1800s. The prominent trench-scar at the top of the Dakota Hogback immediately south and north of the I-70 roadcut is the former Strainland Mine complex (Fig. 79) that operated from approximately 1945 through 1955. South of the Strainland complex is the Parfet fire-clay mine that operated from the early 1900s until 1953. The south Golden fire-clay mines were operated by G.W. Parfet and his heirs, the Denver Brick and Pipe Company, and the Denver Fire Clay Company.[206] A railroad, the Morrison Branch of the Denver and Intermountain Railway Company, operating from 1910 to 1953 (Chapter 12) served these mines.

Porcelain and the Coors Connection

We commonly think of porcelain as fine china and dishware, but grades of dishware range from pottery to fine china. In the 1860s, getting dishware was difficult in the isolated Western U.S., thus a local dishware source could make a profitable business, especially with a railroad connection to Denver beginning in 1870. With modest beginnings, the porcelain industry in Golden became much larger and more important than simply making dishware.

The Herold China and Pottery Company, which was formed in 1910 by ceramist John Herold (1871-1926)[207] and Adolph Coors (1847-1929)[208] as President and investor, used the high-quality Dakota fire-clays from the Golden area. Herold caught Coors' attention after winning recognition at the Denver Keramic Club exhibition at the Brown Palace Hotel. To jumpstart the business, Coors leased its vacant bottle manufacturing plant at 600 8th Street to Herold China at no cost. Through a series of

---

[204] The Colorado State Mine Reclamation Board plans to back-fill these trenches in 2022 to improve safety conditions (Chapter 12).

[205] Like in north Golden, the start-dates of the south Golden clay mines are difficult to determine. Clay, while valuable, was not reported by the Colorado State Bureau of Mines as much as coal and metals, which began official reporting in 1895.

[206] Van Sant, 1959, supplemented by data in Amuedo et al., 1978.

[207] John Herold was a former art superintendent at the prestigious Roseville Pottery plant in Zanesville, OH. Herold departed Golden in 1915 and returned to Ohio, working at the Guernsey Pottery in Cambridge, OH, near Zanesville. That same year Coors legally enjoined Herold from disclosing Coors' porcelain trade secrets to Guernsey Pottery (Kostka, 1973).

[208] Adolph Coors (1847-1929), a German immigrant trained in beer brewing, arrived in Golden in 1873. With majority-partner Jacob Schueler, Coors opened the Golden Brewery in an old tannery purchased from Charles Welch. With a real head for business, Coors bought out Schueler in 1880 and eventually became successful in cement, real estate, mining, malted milk, and ceramics enterprises, but most of all brewing beer. He was a leading citizen of Golden.

recapitalizations, by 1913, Coors became majority owner of Herold China, and his son, Adolph Jr., became General Manager. In 1914, Coors bought out the other stockholders and took full ownership. He continued manufacturing high-quality fine china trademarked as Rosebud China (Fig. 81) and began upgrading the company's technology.[209] In this regard, Coors' connection with CSM was advantageous to the ceramic venture. CSM Professor Herman Fleck helped Herold China perfect its glazing technique, and in 1914, CSM professors attested to the quality of Coors "fireproof china" for industrial laboratory applications.

Fig. 81. Coors Porcelain, Rosebud China on display at the Golden History Museum.

Why was Coors interested in non-beer industries? First, prohibition fervor in Colorado was accelerating through the early 1900s. Coors, being an astute businessman, recognized a need to diversify. In 1914, the State of Colorado enacted a law effective 1 January 1916 prohibiting alcohol sales. With Colorado as an early adopter of what was to become the 18th amendment to the U.S. Constitution in 1919, Coors Brewing had to change gears.

Second, by 1914, world events also created a big difference at Coors. Prior to World War I, most laboratory and industrial ceramics, not to mention domestic bakeware, were imported from Germany. With the outset of WWI in Europe and a 1915 U.S. embargo on German goods, the U.S. Government sought alternative sources for high-quality ceramics.[210] Coors had been upgrading its technology and in 1915 was among seventeen companies that submitted ceramic samples in response to the U.S. request. Only two companies met the specifications and received a contract: Coors (then Herold China) and Champion Spark Plugs. As a result, Coors Porcelain became the dominant national supplier of labware and grew into one of the world's largest porcelain and ceramic businesses. Indeed, during the Prohibition Era (1916–1933), sales of its clay-based products helped keep Coors afloat. In 1920, Herold China became Coors Porcelain, and in 1986 Coors Porcelain became Coors Ceramics. In 2000, the company was renamed to CoorsTek, which flourishes today as a $1.3B leader in the technical ceramics business.[211] Coors is proceeding with plans to re-establish the worldwide CoorsTek headquarters at their landmark business location on 8th and Washington Street. And it all started with Golden fire-clay from the Dakota Group.

Laramie Brick-Clay

The Laramie Formation in and around Golden yielded enormous quantities of clay and still does. Brick-clay contains iron and magnesium mixed with aluminum, making it fire at lower temperatures than aluminum-rich fire-clay; hence its use for brick and pipe-making. Laramie brick-clay mines extended from Clear Creek to south of Alameda Parkway from 1866 through the present (Fig. 78).

The Rubey Mine was the first large-scale Laramie brick-clay mining operation in Golden (Figs. 79 and 82), located on what is now the west side of the CSM campus. Limited brick-clay mining at the Rubey Mine site likely started around 1866.[212] In 1877, G.W. Parfet began large-scale operations. From 1877 until the 1940s, the Rubey Mine was mostly an underground mine, using stope-mining methods. After World War II it became an open-pit quarry until abandonment in 1964, when it was transferred to CSM. The Hoyt Mine was another early Laramie brick-clay mine in Golden, likely beginning in the late

---

[209] Schneider, 1984.
[210] CT 8 February 1912, and Coors Porcelain, 1935.
[211] Beer cans and body armor: Coors jumpstarted the recycling industry by inventing the aluminum beer can at CoorsTek. Also, many military and law enforcement personnel may not know the thanks they owe to CoorsTek for manufacturing ceramic body plates (Cera Shield) used in protective vests, aka flak jackets.
[212] CT 6 November 1867.

1870s. It was abandoned by 1900, but the workings (Figs. 78 and 82) are still visible at the south end of Illinois Street, along the Illinois Street Trail.

Other major Laramie brick-clay mines within Golden (Fig. 82) included the Rockwell Mine in the area now occupied by Fossil Trace Golf Course; the Rubey Standard Mine now under the Jefferson County Laramie building and the Splash Water Park; and the North Apex and Apex (aka Queen City) Mines, occupying a strip of ground from beneath the Golden Terrace Mobile Home Park and stretching south across today's Rooney Road Athletic Complex. These clay mines began operations in the late

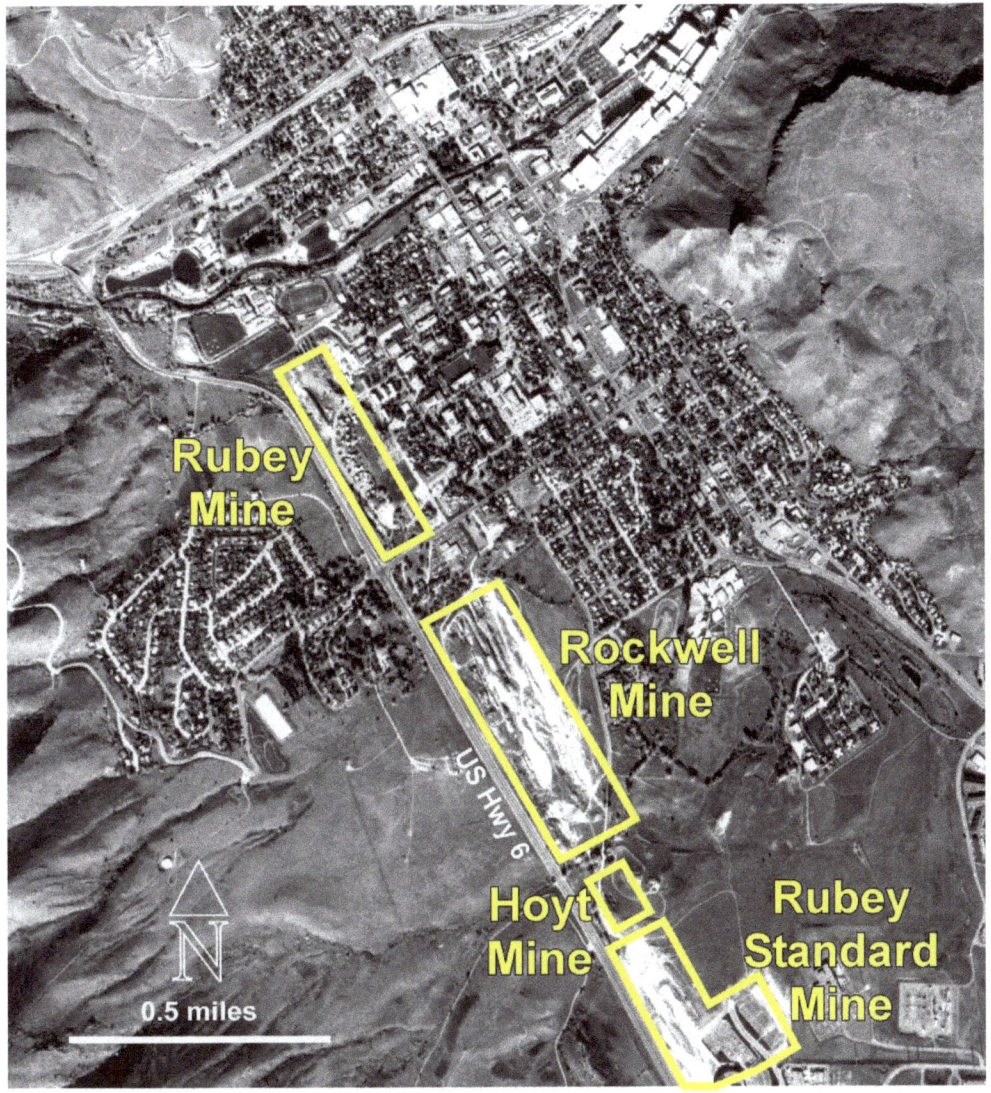

Fig. 82. Laramie clay mines in central Golden showing the active Rockwell and Rubey Standard Mines in September 1990. The Rubey Mine ended operation in 1964 when it was transferred to Colorado School of Mines. The Hoyt Mine ended operations around 1901. Aerial photograph from Colorado Aerial Service, used with permission.

1880s, accelerating in development through the 1920s with underground operations. Mining paused during the Great Depression and resumed after World War II with surface mining. Clay mining in the Rockwell and Rubey Standard Mines continued through the 1990s. The Apex Mines in the Laramie Formation were abandoned by the middle 1950s. The Green Mountain Mine, begun in the 1960s, is now buried under the I-70 to C-470 interchange (Fig. 78). Farther south of Golden, Laramie brick-clay mining extended across the western flank of Green Mountain, following the outcrops of the Laramie Formation.

Active clay mining in the Laramie Formation moved to the southwestern-most flank of Green Mountain at the North and South Chieftain Mines, east of C-470.

Tons of Bricks

A ton of bricks comes from about a ton of brick-clay, without which bricks would not exist. Humble brick-clay has been used for the last 10,000 years to build homes, monuments, and other remarkable buildings. Beginning 6000 years ago, it has been used to make pottery and, yes, sewer pipe, i.e., plumbing, essential for public health and welfare.

Bricks were a much-desired building material in a place where the native vegetation included no trees. Adding to a lack of timber, in April 1863, bricks became in very high demand due to a major Denver City fire, which led to a Denver ordinance requiring all new major buildings to be built of brick. It was then that Golden became the center of brickmaking in northeastern Colorado with its wide variety of local clay and coal to fire the brick kilns. Brickworks established after 1864 included the Golden City Pottery and Fire Brick Works[213] (1866 by WAH Loveland); Golden Pressed Brick (originally the Rocky Mountain Fire Brick and Tile Company, 1873); Chicago Pressed Bricks (1879); and the Cambria Tile and Brickworks (1879).

Fig. 83. The Golden Pressed Brick Company yard in north Golden circa 1940. View west to Mount Galbraith. Argall, (1949, Figure 9, p. 100 in the CSM Quarterly, v. 44, no. 2), used with permission.

The Golden Pressed Brick Company was by far the largest brickworks in the 1800s. Founded in 1873 by Jacob Koenig and Fritz Fischer as Rocky Mountain Fire Brick and Tile Company, it was located along the Colorado Central rail line at what is now 1122 8th Street, near today's Lookout Landing Condos. In 1890 the company was acquired by John B. and William Church with co-investor G.W. Parfet, who renamed company as the Golden Pressed Brickworks. In 1893, they added a second site north of Golden Gate Canyon road between today's Brickyard Road and Catamount Drive. In January 1895, the 8th Street brickworks burned,[214] and the Church brothers concentrated on their northern brickworks, the Golden Fire-Brick Company (Fig. 83), which was twice expanded and finally shut down in 1963. The Brickyard Superintendent's House (Fig. 84) is one of the last remaining brickyard structures in Golden, located on the east side of Catamount Road in north Golden. It is owned by the City and leased to the Golden Landmarks Association.[215]

Fig. 84. Brickyard superintendent's house in north Golden, east of Catamount Street in 2019.

Golden's brick legacy lives on all over Colorado. Golden bricks are in most buildings in Golden and Denver as well as across the State.

---

[213] It was never rebuilt.
[214] CT 12 April 1893, and Gardner, 2017. This became the Golden Fire Brick Company.
[215] http://goldenlandmarks.com/brickyard-house/ and http://goldenlandmarks.com/brickyard-history/ give the history of this building and the surrounding brickworks from the Golden Landmarks Association

The Parfet Connection

Brick-clay mining was the domain of the Parfet familty for 124 years beginning with George (G.W.) Parfet (1859-1924) in 1877 and ending with his heirs in 2002. At the ripe age of 19 years old, G.W. Parfet, Sr. took his knowledge of underground coal mining and applied it to clay mining on land in the Laramie Formation at the end of today's 12th Street. Initially called the Parfet #1 or the White Ash Clay Mine, it eventually became known as the Rubey Mine due to financing from H.M. Rubey and W.S. Woods on land leased from the Colorado Limited Company owned by W.A.H. Loveland. After mining coal for others as a teen, Parfet now had a clay mine of his own.[216] Not limiting himself to brick-clay, G.W. Parfet's "empire" soon included fire-clay mines. He operated the Santa Fe Mine in the Dakota Formation along Eagle Ridge, and the Apex 10, South Golden, and Parfet mines on the Dakota Hogback between Lena Gulch and south of I-70 (Fig. 79). Principal customers were the Cambria Brick and Tile Company in Golden (13th and East Street) and the Denver Sewer Pipe and Brick Company. Continuing his legacy, his son (G.W. Parfet, Jr., 1889-1940), grandson (William "Bill" G. Parfet, 1918-1998), and great-grandson (William "Chip" G. Parfet, Jr. 1946-present) operated four Laramie brick-clay mines in addition to the Rubey Mine: the Rockwell, Rubey Standard, Apex and North Apex, and Green Mountain Mines (Fig. 79).

Fig. 85. Plaque at the 11th Hole on Fossil Trace Golf Course commemorating Parfet and partners' donation of the clay pits to the City of Golden.

While clay from the Parfet-family mines in south Golden was mined out by 1953,[217] the family continued clay mining in the Laramie Formation in central Golden until August 2002, a mining legacy of 125 years (1877-2002)!

Over time, the many Parfet-family mine-sites were developed for housing and government buildings or reclaimed for recreation and open space. The worked-out Rubey Mine was deeded to the School of Mines in 1964 in exchange for clay-bearing acreage further south. The Rockwell Mine was donated to the City of Golden in 2002 to be used for part of the Fossil Trace Golf Course (Fig. 85). Part of the Rubey Standard Mine was sold to Jefferson County as the site for the Laramie Building, completed in 2005. Another worked-out part of the Rubey Standard Mine became The Splash Water Park, owned by the City of Golden. The North Apex Mine was developed as a mobile-home park, whereas the Apex Mine became a Jefferson County landfill, operated from 1958 to 1980. In 2007 the landfill was reclaimed for the Rooney Road Athletic Complex.

---

[216] The Golden Kiwanis Club established Parfet Park in 1929 on the southeast corner of Washington Avenue and 10th Street in honor of its first president George W. Parfet, Sr. Parfet Park is the oldest park in Golden and is also the site of the Boston Company's 1859 first log building in Golden. Also, as a major environmental and olfactory improvement, the Park replaced the town dump.

[217] Bill (William G.) Parfet recounted the post-1940 clay operations in a 1977 interview by Bill Ryder of the Colorado Railroad Museum (Parfet and Ryder, 1977).

Chapter 11: No Stone Unturned

Golden has been a center for stone and aggregate mining since its founding in 1859, an activity that accelerated with road building in the early 1900s and then with urban growth in the 1950s and 1960s. Over the years, ten major stone and aggregate quarries existed in Golden (Fig. 86): sand and

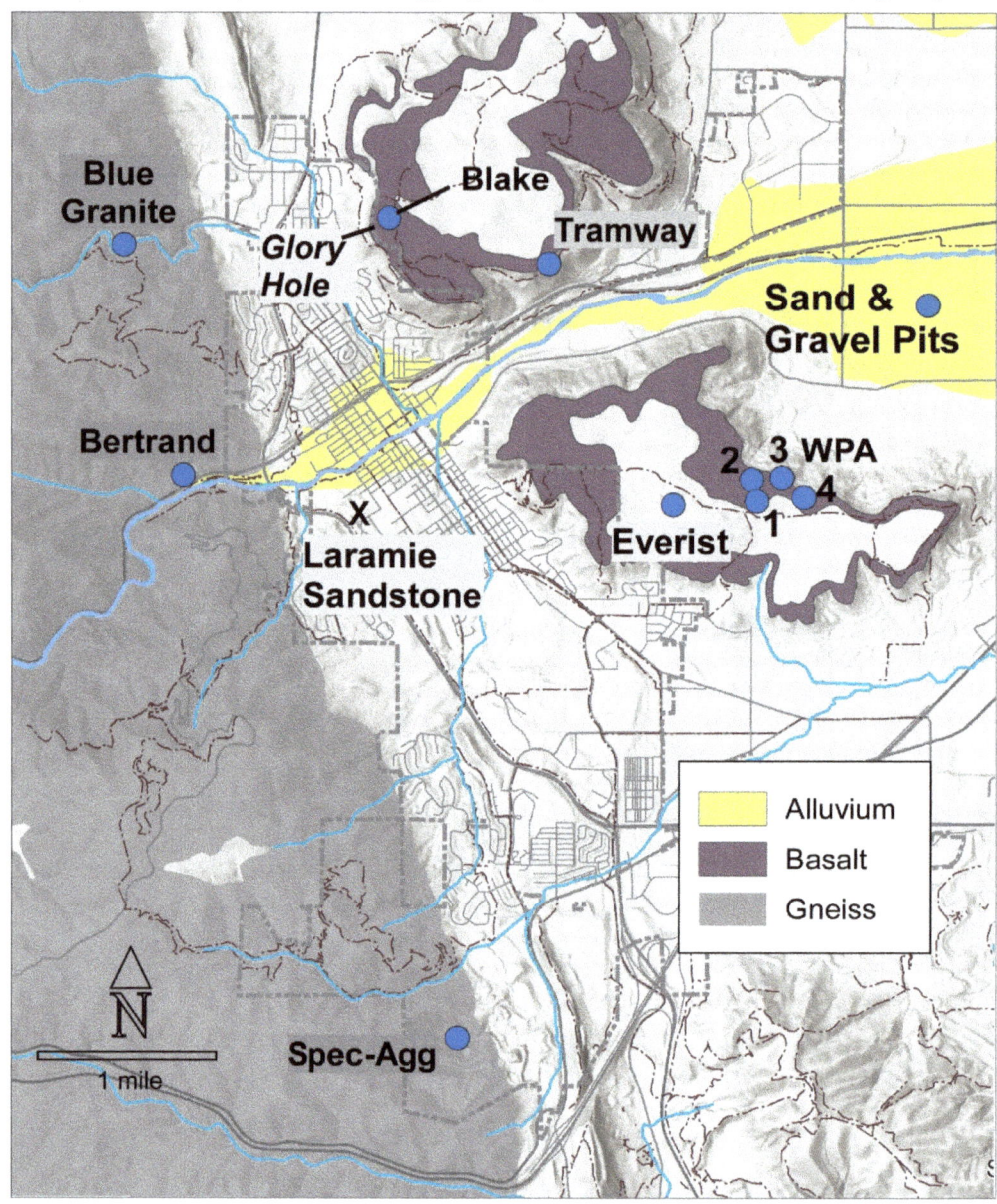

Fig. 86. Historic and present-day crushed rock and building-stone quarries. A minor quarry in sandstone of the Laramie Formation (X) operated for a few years in the 1860s. Of these, only the Spec-Agg Quarry is active today. Based on Schwochow et al. (1974), Butler (1992), Gardner (2017), and aerial photographs.

gravel quarries in the Quaternary alluvium along Clear Creek; gneiss in the Precambrian bedrock along the Front Range; basalt capping the two Table Mountains; and, for a short time, in sandstone of the Laramie Formation. As Golden grew from a tent camp to a proper town, Euro-American settlers initially used native rock to build permanent structures by quarrying it from nearby outcrops. For example, the Astor House (1867)[218] was built from sandstone blocks quarried from the Laramie Formation at the west

---

[218] History Colorado, National and State Register of Historic Places Inventory #PH0087033.

end of 12th Street.[219] This quarry was later the site of the large Rubey Mine for clay (Chapter 10). Rock from the small, now-abandoned Blue Granite quarry near the mouth of Golden Gate Canyon (Fig. 86)[220] formed the walls of many buildings in Golden, although none of those buildings remains today.[221] And, in one case, boulders from alluvium in Clear Creek are in the walls of the Armory Building (1913, Chapter 6). But by 1870, the rise of the brick industry in Golden (Chapter 10) replaced the use of native stone for buildings. Limestone was another early commodity used for making mortar, the "glue" that held bricks and building stone together. After 1900, the rise of the automobile and demand for surfaced streets[222] created a large demand for aggregate (sand, gravel, and crushed rock), a major component of surfacing material, as well as concrete, riprap for stabilizing embankments, and landscaping material. Perhaps not as alluring as gold and other precious metals, the non-energy mineral[223] resources of crushed rock, sand, and gravel together comprise invaluable commodities that we take for granted, just like clean drinking water.

The Bastard Limestone

The Glennon and Falcon members (the "bastard limestone" of Hayden in 1869) of the Lykins Formation (Chapter 2) were quarried for limestone in Golden between 1859 and 1898.[224] Within Golden city limits, remaining shallow trenches where limestone was quarried from vertical beds are on the CSM survey field (Fig. 87). Other trenches and ruins of associated kilns exist along the Lykins Formation outcrop between US Hwy-285 and I-70 through Morrison and north of Golden along Pine Ridge Road to Ralston Creek. Quarried limestone was crushed and then heated to 2000°F in kilns to produce quicklime (CaO or calcium oxide), a bright white substance. Quicklime mixed with water produced a substance called slaked, or hydrated, lime (Ca(OH)$_2$, or calcium hydroxide).

Fig. 87. Linear limestone trench (arrow) on CSM survey field. White patch on right is a pile of quicklime. View to north.

Lime used for mortar was critical to building brick structures. As such, Golden was a thriving but short-lived center for the quicklime industry in the late 1800s. By October 1898,[225] Golden only had one operating lime kiln within the city limits. Two years later, no operating lime kilns remained, and they fell into disrepair. Competition from the invention of Portland cement and gypsum plaster derailed the lime industry in the Golden area.

Cambria Lime Kiln

The historic Cambria Lime Kiln (Fig. 88) along Kinney Run Gulch is the only <u>restored</u> lime-kiln in Jefferson County. It was built in 1879[226] by John Hodges as the first structure of his Cambria Brick and

---

[219] Gardner, 2017, p. 25, and gardnerhistory@aol.com
[220] It is not a granite. Rather it is a hornblende-plagioclase gneiss according to Kellogg et al. (2008), but who is counting?
[221] Simmons et al., 2013. The quarry is still visible in Golden Gate Canyon near the Mt. Galbraith trailhead.
[222] Street surfacing first began with paving stones and crushed and packed-down rock, then to concrete. Asphalt paving was used only after the mid-1920s, because asphalt, derived from oil was not developed in Colorado until after the mid-1920s.
[223] Non-energy mineral resources, also referred to as industrial minerals, include clay, aggregate, building stone, limestone, gypsum, and potash. By contrast, <u>energy</u>-mineral resources include coal, oil, and natural gas.
[224] Lewis, 2013.
[225] CT 26 October 1898.
[226] CT 18 June, and 16 July 1879.

Tile Company, then located at 13th and East Streets (now a Coors Brewery parking lot). Charles Welch[227] was superintendent for the Cambria Brick and Tile Company. Welch conveniently co-owned the lime beds leased to the Cambria Company and was the President of the Golden City and South Platte Railroad

Fig. 88. Cambria Kiln along Kinney Run Gulch in the early 1900s (left), from the Gardner Family Collection, and partially restored as in a 2021 original painting by Jason Slowinksi (right), both images used with permission.

(GC&SP). With that key business connection, Welch built a GC&SP branch to the new kiln in order to transport lime back to the Cambria Brickworks (Chapter 12). Coal used to fire the Cambria Lime Kiln and the brick kilns at Cambria Brickworks[228] was likely mined at the nearby Johnson coal mine on Hoyt's Ranch at the south end of today's Illinois Street (Fig. 72, Chapter 9).

The Cambria Kiln was built with Lyons Formation sandstone from the cliff behind the kiln, on the east side of Kinney Run Trail. It was conveniently located near the "lime seams" in the Glennon limestone, also east of the Kinney Run Trail (e.g., Fig. 79, Chapter 5). The kiln operated until 1892, when exhaustion of the local lime seams and the advent of cheaper and better Portland cement sidetracked the lime/mortar business in Golden. In 2005, a City Council Resolution recognized "the unique geologic areas in the Eagle Ridge area, including the Dakota hog back, Cambria Lime Kiln, Kinney Run Trail, and riparian areas," and designated them "as a geologic and environmental education park." In 2009, the City of Golden and the Golden Civic Foundation restored the base of the kiln and placed interpretive signs.

Sand and Gravel Quarries

Quaternary alluvial (sand, gravel, and cobble) deposits along Clear Creek, especially those east of the Ford Street Bridge in Golden, were of exceptional quality for high-grade concrete, gravel, and sand.[229] Quarried extensively through the 1970s, the deposits are now abandoned due to urbanization, with some pits now filled in to provide building sites. East of the Coors Brewery, several mined-out water-filled pits (Fig. 86, "Sand & Gravel Pits") are now clay-lined and used for industrial processing water. A couple of unlined pits are "reservoirs" supplied with water from the Wannemaker and Rocky Mountains Ditches, both owned by Coors, and used for brewing water.[230]

---

[227] Charles Welch, Senior (1830-1908) had his fingers in just about everything in Golden in the mid to late 1800s, along with WAH Loveland, his business partner.
[228] That same GC&SP branch likely transported coal, as well as lime, back the brickyard. See Chapter 12 for historic rail lines.
[229] Schwochow et al., 1974.
[230] Arbrogast et al., 2002, p. 12: includes an excellent history of sand and gravel operations and land use changes along Clear Creek east of Golden.

Gneiss Quarries

Precambrian gneiss in the Front Range west of Golden and at other locations along the Front Range is a source of crushed-rock aggregate. In Golden, two gneiss quarries have had different histories. One still operates, and the other is abandoned.

Fig. 89. The Specification Aggregate (Spec-Agg) Quarry in 2019. View to west from the south Dakota (Tincup) Hogback.

The Spec-Agg Quarry (formally the Specification-Aggregate Quarry) south of Apex Canyon and west of Colfax/US Hwy-40 (Figs. 86 and 89) began in 1965 as the Holloway Quarry, producing the riprap used in constructing Chatfield Dam.[231] Since then, the quarry has changed hands and expanded several times. Owner/operator from 1970 to 2011, LaFarge N.A. successfully enlarged the quarry in 1972 before it was annexed by the City of Golden. In 2003, the quarry was enlarged to its current 285-acre footprint via a land swap. LaFarge N.A. transferred corporate property totaling over 400 acres, including the former Blake Quarry on North Table Mountain, to Jefferson County Open Space in exchange for 60 acres adjacent to the existing quarry. The expansion extended the life of the quarry by at least 20 years and provided space for a ready-mix concrete plant to supplement an existing asphalt plant.[232] The exchange also established the North Table Mountain Park, much to the enjoyment of the residents of Golden and the Denver metropolitan area. Today the Spec-Agg Quarry is operated by Martin-Marietta Aggregates (MMA). In 2019, MMA proposed another expansion, to add 64 acres to the south side of the quarry in exchange for conveying Heritage Square and Bachman Ranch property to the City of Golden and Jefferson County Open Space. As of 2021 the proposal was still under consideration and the subject of public hearings.

Fig. 90. Former Bertrand quarry in Clear Creek Canyon in 2020. View to north.

The Bertrand Quarry[233] was located ½ mile west of the mouth of Clear Creek Canyon along the north side of today's US Hwy-6 (Fig. 90). Opened in 1926, it was enlarged several times until operations ended in 1973.[234] In 1949,

---

[231] Schwochow, 1980, p. 38.
[232] Langer and Tucker, 2003, p.8.
[233] Past names for the Bertrand Quarry include Cass and Goltra.
[234] Brannon Sand and Gravel Company, 1979.

rock from the quarry supplied riprap for Bonny Dam in eastern Colorado.[235] A large blast from the quarry in October 1949 rattled the homes of Golden residents, some of whom sued for damage and received a settlement.[236] In the 1960s, with increased demand for paving roads, I-70 in particular, the quarry was back in operation with restricted blasting. The 1973 variance granted to Asphalt Paving by Jefferson County to continue mining was allowed to expire. A plan to renew quarrying in 1978 was allowed to lapse due to local opposition and concerns about slope stability. However, in the 1980s Brannon Sand and Gravel proposed a large quarry on land that they owned around Mt. Galbraith, with access from the former Bertrand Quarry. The proposal met with strong opposition, much of which was focussed on truck traffic. In 1995, the Brannon acreage was purchased by Jeffco Open Space to become Mount Galbraith Park. An 80-acre land trade with O.C. Goltra in 1997 put the former Bertrand Quarry site into Jeffco Open Space. In 2020, Jeffco Open Space used stockpiled crushed rock at the old Bertrand site for construction of a new trailhead for the Clear Creek Open Space Park, thus reducing rock-fall hazard.

Basalt Quarries

Since 1899, North and South Table Mountains have been a source of crushed-rock and building stone mined from basalt-lava Flows 3 and 4 that cap the mountains (Chapter 4). As the need for crushed-rock aggregate for railroad ballast and road-paving material arose in the late 1900s, quarrying began and then accelerated through the 1920s on North Table Mountain and through the 1930s on South Table Mountain. After World War II, basalt was in demand for riprap for dams being constructed in the Denver area and on the Plains to the east. During the rapid urbanization of Denver in the late 1960s, demand for aggregate again increased, leading to more quarry operations on North Table Mountain and more quarry proposals on South Table Mountain. These activities collided with changing public attitudes about mining and the environment in the early 1970s.

North Table Mountain: From Zeolites to Riprap

The top of North Table Mountain, now all part of the North Table Mountain Open Space Park, was the site of two basalt quarries. Basalt-lava Flows 3 and 4 were quarried from the mid-1870s through the 1970s. The earliest record of a quarry on North Table Mountain was 1874 when a small quarry was opened by Arthur Lakes and his CSM students for the purpose of obtaining collector-prized zeolite minerals (Chapter 4). Quarrying for paving stones and aggregate, however, began in 1899 by F.O. Blake, owner of the Blake Asphalt and Paving Company.[237]

Fig. 91. A. Tombstone made of North Table Mountain basalt in the Golden Cemetery. B. Basalt paving stones at Tramway Quarry. Scoops (S, arrows) in stones accommodated streetcar rails. Horizontal groove (G) in bottom stone is from quarrying the rock.

The oldest quarry on North Table Mountain, eventually called the Tramway Quarry, occupies the site of the former zeolite quarry opened by Arthur Lakes in 1874 on the south side of North Table Mountain (Fig. 86). Afterward, in 1901, the mineral-collecting quarry became the second of two Blake quarries on North Table Mountain. The massive basalt cliffs forming the rimrock were quarried for "monument" stones (tombstones, Fig. 91),

---

[235] CT 10 November 1949.
[236] CT 15 October 1949 and 2 April 1953.
[237] Blake opened two quarries on North Table Mountain, both initially called the Whenestone quarries (CT 18 September 1901). We use the name Tramway Quarry, for the south-side quarry, which was the former site of Arthur Lakes' zeolite quarry.

paving stones, and crushed rock. Piles of unused paving stones remain at the site today (Fig. 91). Blake's biggest customer was the Denver City Tramway Company (later the Denver Tramway Company), which had a substantial program of building roads and maintaining electrified interurban (streetcar) rail lines in the Denver-Golden area.[238]

The Denver Tramway Company (DTC) took over operations of the Tramway Quarry in 1905. One of the difficulties of quarrying 700 feet above the Clear Creek Valley was getting the quarried stone down the south side of the mountain. After some near-miss disasters with runaway wagons, the DTC installed an inclined tram in 1907[239] to transport rock more safely and quickly to a crushing facility in Clear Creek Valley, next to a rail line (the 83-Line, see Chapter 12). Worked on-and-off, the quarry was rejuvenated to make a gravel-paved road for West Colfax and South Golden Road in 1917. But by 1923, the Tramway Quarry was abandoned, due to bankruptcy of the DTC and to competition from the other quarry on North Table Mountains, the Blake Quarry. The Tramway Quarry never re-opened after 1923.

On the west side of North Table Mountain, the first Blake Quarry[240] was opened in 1899 by F.O. Blake,[241] who had leased the entire top of North Table Mountain with the intent of quarrying the thick basalt flows, present everywhere across the mountain. But realistically, quarrying focussed on the edges of the mountain, because getting the rock down the hill, like at the Tramway Quarry, was a big issue in the days of horse-drawn wagons. In 1909, the Wescott-Doane Company took over operations from Blake.[242] They greatly increased the capacity of the Blake Quarry with a new, all-electric crushing facility at the base of the mountain. They also installed a gravity-driven tramway down the mountainside to a branch of the Colorado and Southern (C&S) Railroad in order to transport the rock (the Brickyard Branch, see Chapter 12).

Around 1924, the Gibbons and Lawrence Construction Company opened a two-acre pit, enlarging the former Wescott-Doane operation, and quarrying nearly 2000 tons per day for use as concrete aggregate. They also dug a deep pit, the Glory Hole (Figs. 86 and 92), between the main quarry and the cliff face as part of a rejuvenated gravity-tram operation. The 100-foot-deep Glory Hole took large pieces of basalt quarried from the mountain top and basically dropped (that is the "gravity" component) them into the hole and through a tunnel at the base of the cliffs, where rock was then loaded into the gravity-tram and taken down the mountainside to a crusher (Chapter 12).[243] Tragically, Lawrence's partner, Eddie Gibbons, was crushed by a 50-ton rock-slide in the Glory Hole on 30 June 1924. Again, let's just say operational safety in 1924 was not the same as today's standards.[244] As another part of the operation, a steep, ½-mile-long road was built from the base to top of North

Fig. 92. Former Blake Quarry and the Glory Hole (arrow) on the west side of North Table Mountain in 2020. View to southwest.

---

[238] CT 18 September 1901
[239] Robertson and Forrest, 2010, p. 79, and CT 25 April 1907, but it was NOT an aerial tramway as some sources report. Rather it had a railbed and trestle configuration (Chapter 12).
[240] The quarry has also been called the Blake, Doane-Westcott, Doane, Rogers, Foss, Rocky Mountain Aggregate, and Western Paving quarry through the years. Here we use Blake Quarry.
[241] CT 29 November 1899.
[242] CT 1 April 1909.
[243] CT 5 February 1925.
[244] CT 3 July 1924.

Table Mountain, providing vehicular access.[245] That road now is the part of the West Trailhead access to the top of North Table Mountain (e.g., Fig. 31 in Chapter 4).

In 1949 the Blake Quarry was expanded to seven acres by the Ralph Rogers Company of Indianapolis. They built a new crushing plant (the site of the current West Trailhead parking lot) with a capacity of 3000 tons per day, becoming the second largest crushed-rock production tonnage in Colorado at the time.[246] In addition to its use as road-paving material, crushed basalt was shipped via railroad as far as McCook, Nebraska to serve as aggregate and riprap for the Harlan County Dam.[247] Between 1958 and 1960, other contractors operated the site for short times. The most recent mining activity in the Blake Quarry was in 1976 when Western Paving Construction Company processed some of the rock piles left by earlier operations, using the steep road for transport down the mountain.[248] That short-lived operation was contentious with Golden residents due to noise, dust, and truck traffic, and it was discontinued after 1976. The remaining history of quarrying forms part of the movement to acquire the land on and around North Table Mountain for Open Space, a story that is part of the mining legacy of Golden (Chapter 12).

South Table Mountain: The WPA Quarries

On top of South Table Mountain, aggregate was quarried intermittently from about 1900 through the early 1970s, also from basalt-lava Flows 3 and 4.[249] Earliest references to quarrying are in 1900 when 3500 tons of basalt were quarried by F.O. Blake, "just east" of Castle Rock.[250] Another early reference to quarrying was that of Thomas B. Doane around 1910, at first to provide ballast for the roadbed of the Denver and Intermountain Railway.[251] Possibly both references were to areas occupied by the later WPA quarries, as evidence for neither is obvious from aerial photos taken in the early 1930s.

Plans to quarry basalt on South Table Mountain were revived in 1917 by the Quaintance Investment Company, which owned the land including Castle Rock. The company intended to use the funicular railway built in 1913 to move the quarried stone down the hill. For better or worse, those plans never materialized. Instead, the company improved the "Lava Lane" dance hall on top of Castle Rock, then a meeting place for the Ku Klux Klan (KKK) organization in the Denver area, until its demise by fire

Fig. 93. Former WPA Quarry #2 on the east side of South Table Mountain. Cliff of quarry face is over 40 feet high. View east.

---

[245] Argall, 1949, p. 124.
[246] Ibid. p. 124.
[247] Ibid. p. 125.
[248] Schwochow, 1978, unpublished.
[249] Schwochow, 1980, p. 37
[250] CT 13 June 1900 and 27 June 1900. Probably near the later WPA quarries on the east side of the mountain.
[251] CT 4 February 1909.

on 7 August 1927.[252] Of note, however, an automobile road, also called Lava Lane, was constructed from Quaker Street on the south side of South Table Mountain and north across to Castle Rock in 1920, to great acclaim by those who drove that road.

A large quarry complex, herein termed WPA Quarries 1-4[253] (Figs. 86 and 93), was established during the Great Depression in 1934, initially as a Federal Emergency Relief Administration project that was transferred to a Works Progress Administration (WPA) project in 1935.[254] Then, as part of Camp George West, the quarry operation hired idled and homeless men when the unemployment rate in Colorado was about 25%.[255] Employing 350 men at start up, dynamite was used to quarry the basalt, which was then crushed onsite and removed by trucks via Quaker Street. The mining access road to the WPA Quarries, the former Lava Lane, began at the end of Quaker Street on the south side of the mountain, near what is now the National Renewable Energy Laboratory (NREL), and now is called the Old Quarry Trail. By July 1937 nearly 200 truck drivers hauled the material to various projects around the region, including paving Alameda Parkway across the Dakota Hogback and roads around Red Rocks Park near Morrison, and providing riprap for stabilizing the banks of the South Platte River and Cherry Creek. Ownership of the 80-acre WPA quarry complex was conveyed to the City and County of Denver in 1938.

Fig. 94. Got holes? Drill holes in basalt cliff face of Quarry #2 left behind from testing in the 1960s. View to east.

After a pause during World War II, the Phelps-Wunderlich and James Company operated the quarry, producing 900 to 1000 tons of rock a day for riprap for Cherry Creek Dam in Denver.[256] Quarries #1 and #2 also served as a rock-mechanics test-sites operated by the Gardner-Denver Company for a short time in the 1960s. That company, in collaboration with the Colorado School of Mines Research Institute (CSMRI, Chapter 8), used the rock faces to test the engineering design of rock bolts,[257] leaving behind a myriad of drill-holes arranged in rectangular patterns across the quarry faces (Fig. 94). Quarry #4 became the Table Mountain Gun Club that operated from the early 1950s through the early 1970s.[258] A part of the WPA Quarry complex became the South Table Mountain Open Space Park, formally established in 2002. The remaining quarry areas are private property, owned by Bear Creek Development Coporation, the latest successor to the Quaintance Investment Company.

A 32-acre Everist Quarry (Fig. 86) was a short-lived scrape on the land in 1974. According to newspaper articles,[259] the L.G. Everist Company acquired a variance from Jefferson County in 1970 to mine for crushed-rock on top of South Table Mountain on land owned by Coors, and using the Quaker Street access (Lava Lane) from the WPA years. Conditions of the Jefferson County permit included paving Quaker Street before mining. In May 1974, residents in the area noticed that quarrying

---

[252] CT 11 August 1927. The dance hall had been leased to the KKK by owner Charles Quaintance, one of the few KKK members in Golden (History Colorado KKK Ledger, 2021, and Denver Post, 2021). The hall was unused for more than a year before the suspicious fire. By 1927, KKK influence was waning into obscurity due to many scandals (Colorado Encyclopedia, 2021).
[253] Numbered as in Butler, 1992, Figure 24.
[254] Ibid. p.11.
[255] US Dept. of Interior, 1992, National Park Service National register of historic places: Camp George West.
[256] Ibid. and Argall, 1949, Table 44.
[257] CT 18 August 1966.
[258] CT 2 February 1972.
[259] CT 16 June 1970, 9 August 1973, 4 October 1973, and 8 May 1974.

operations had begun, but Quaker Street had not been paved. As it turned out, a mining permit had not been issued by the Colorado State Land Reclamation Board, although the operator maintained that all needed permits had been acquired. Operations ceased. Finally in 1994, after two decades of increasingly contentious quarry and development proposals, Molson-Coors sold the Coors-owned portions of South Table Mountain to Jeffco Open Space, including the Everist Quarry and the mining access road from Quaker Street. The lands owned by Jeffco Open Space became South Table Mountain Park.

Chapter 12: Mining Legacies

Mining forms a large part of the Euro-American history of Golden as well as the rest of Colorado. In Chapter 7, we highlighted the legacy of mine-processing operations and water pollution. Here, we feature mining-related historic rail operations in Golden, abandoned mine subsidence, and open space. Abandoned mines create modern construction and attractive-nuisance hazards. Also, you might be surprised to find out that much of Golden's open space exists due to former and proposed mines. That story is rooted in changing attitudes about mining and the environment arising in the late 1960s.

Mining-Related Historic Rail Lines

Railroads are intimately linked to the mining history of Golden, from transporting ores from the mountains to Golden to the transport of limestone, coal, and clay within Golden and on to Denver. After the initial rail connection from Denver to Golden was completed in 1870 (CCRR Main Line, Fig. 95), the first period of rail building concentrated on connections from Golden to the mountains and around

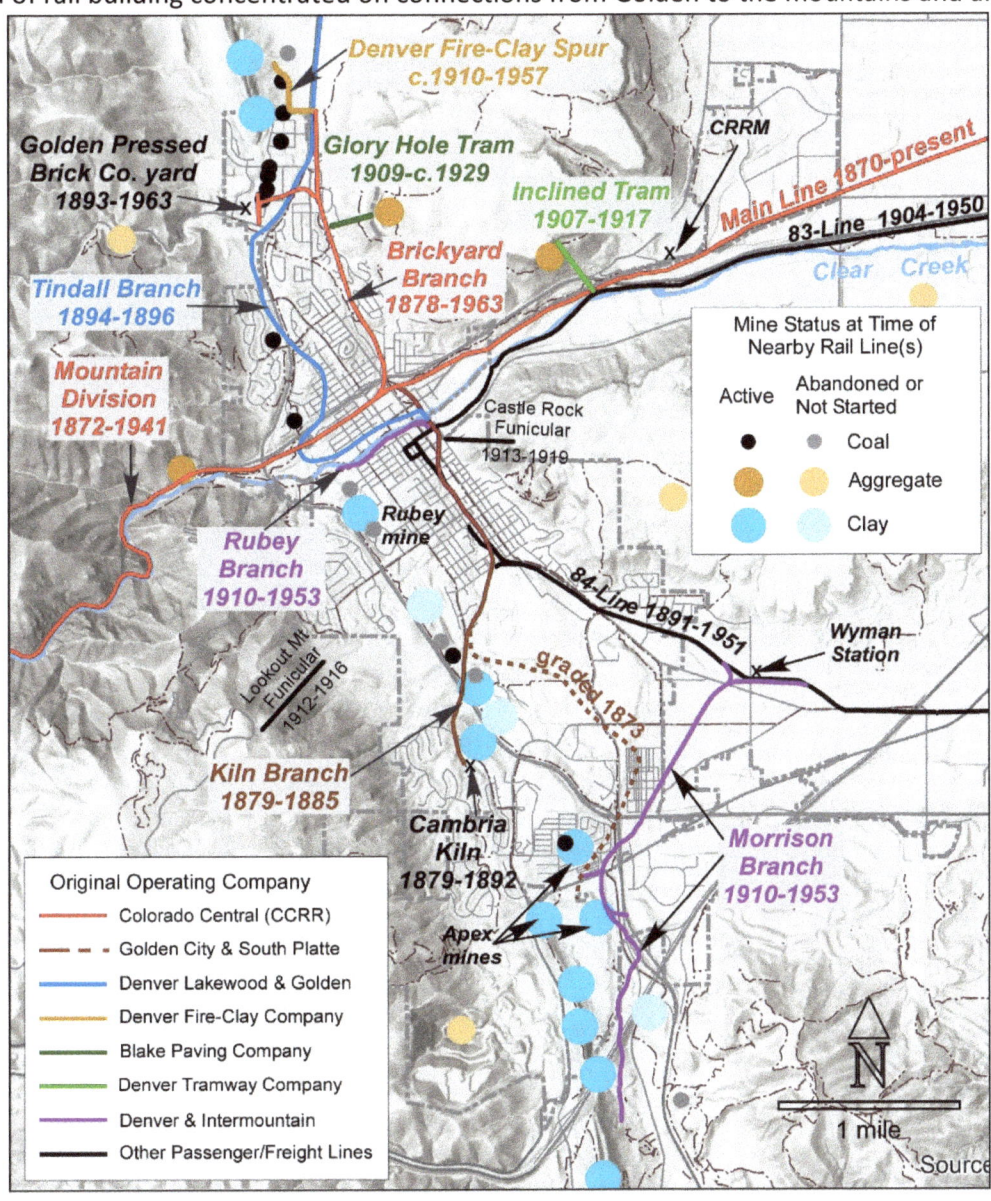

Fig. 95. Historic rail lines and related mines in the Golden area. Rail lines compiled from topographic maps, aerial photographs, Yehle (2001), Robertson and Forrest (2010), Abbott et al. (2007), Willits (1878, 1899) maps, Jefferson County, Colorado archives map, circa 1879. CRRM is location of the Colorado Railroad Museum.

Golden for local uses. The second period focussed on connections to local aggregate quarries and clay mines. From 1872 to 1963, Golden was the site of at least nine mining-related rail lines, including two inclined trams.

The first mining-related rail line was that of the Colorado Central (CCRR) Mountain Division line connecting Golden to the gold mines of Idaho Springs and Black Hawk. Constructed by W.A.H. Loveland, Charles C. Welch, and Edward L. Berthoud,[260] that historic narrow gauge line wound its way west up Clear Creek Canyon. Its 1872 completion to Black Hawk, and later to Idaho Springs, Central City, and Georgetown, made Golden a smelting center for about 20 years, that in turn spurred demand for local Golden coal and fire-brick clay for smelter furnaces. Later, when demand for paved roads increased around 1900, the line directly served the Bertrand aggregate quarry near the mouth of Clear Creek Canyon (Fig. 95).

A second mining-related rail line was the CCRR (that later became the Colorado and Southern C&S) Brickyard Branch up Tucker Gulch that began serving the Church's Golden Pressed Brickyard in 1889. It was originally a main line of the CCRR built in 1878 linking Golden directly to Cheyenne, Wyoming.[261] But, abandonment of that route north of Table Mountain[262] turned the track along Tucker Gulch into a branch to the brickyard, later serving the Blake Quarry and successor operating companies. The Denver Fire-Clay Company built and maintained its own spur connecting to the Brickyard Branch until 1957. The railroad company removed most of the rails of the Brickyard Branch in 1963. Urban development since then covered most of the rail bed. However, a portion of the rail bed and its rail is visible along the east side of Tucker Gulch trail, north of 1st Street (Fig. 96).

Fig. 96. Remaining tracks and rail bed along Tucker Gulch north of 1st Street along Brickyard Branch on west side of North Table Mountain. View to south.

With rail lines operating north and west of Golden, Loveland, Welch, and Berthoud saw a need for rail lines to go south from Golden to connect directly with the Denver and Rio Grande Railroad. In 1873, they formed a railroad called the Golden City and South Platte (GC&SP). After spending time and money to grade a line almost to Littleton, CO, as well as construct a trestle-bridge across Clear Creek, the Financial Panic of 1873 stalled further construction for nearly five years. In 1879, the GC&SP finally laid rail to create the Kiln Branch from the railyard in Golden, across Clear Creek, along today's Jackson Street and southwestward up Kinney Run Gulch to the Johnson coal mine, Hoyt brick-clay and Santa Fe fire-clay mines, and the Cambria Lime Kiln (Fig. 95). However, dubious financing and a corporate takeover fight between Loveland and the Union Pacific made the Kiln Branch sub-commercial. The Kiln Branch was abandoned in 1885.[263] Historians argue whether rails were ever laid along the graded 1873 GC&SP rail route leading south to the clay mines in the Apex area where the

---

[260] These gentlemen had a one-track mind: make Golden the center of politics and railroads. Golden was the Territorial Capitol from 1862 until 1867, when Denver "snatched" the designation from Golden on 9 December 1867 after a suspicious vote by the Legislative Council three days earlier. Colorado became a state in 1876.
[261] Grunska, 2003, provides an entertaining, award-winning article on the railroads around Golden and Morrison, CO.
[262] At that point in time, the CCRR, after several mergers and acquisitions, as we call them today, became part of the Colorado and Southern, or C&S. Today's successor company is the Burlington Northern and Santa Fe Railroad (BNSF).
[263] Abbott et al., 2007, p. 38-39. Details on when the tracks were removed are unclear.

Rooney Road Athletic Complex now resides (Fig. 95).[264] If laid, the rails were removed by 1888, as they are not shown on the USGS topographic survey of that year (Emmons et al., 1896, Plate 3).

As interurban, or streetcar, rail lines mulitpled in Denver in the 1880s,[265] investors including Loveland, Welch, Berthoud, and Samuel Newhouse hopped into action, forming the Denver Lakewood and Golden (DL&G) railroad in 1890. The successor to the GC&SP, their intent was to bring electrified streetcar lines to Golden.[266] After building the 84-Line to Denver in 1891, they opened the Tindall Branch in 1894 (Fig. 95), using the former GC&SP trestle-bridge across Clear Creek at the northern extension of today's East Street. With the Tindall Branch, the investors sought to bolster the company's shaky finances by transporting valuable coal from the Tindall Coal Mine along Ralston Creek, north of Golden. Branch operations also tried to capture business from the Church brothers' Golden Pressed Brickyard and the Denver Fire-Clay Mine, directly competing with the C&S Brickyard Branch. Alas, the investors could not have formed a company at a worse time. Under-financing was followed by the Financial Panic of 1893. Then, Mother Nature stepped in with two successive floods on Clear Creek (1894 and 1896, Chapter 6) that twice destroyed the trestle-bridge (Fig. 97). The company went into receivership on July 31, 1896, a week after the 1896 flood. The Tindall Branch lay stranded with no bridge. The railroad company removed the tracks in 1904.[267]

Fig. 97. 1894 view of destroyed DL&G trestle-bridge of the Tindall Branch after flood on Clear Creek. View to southeast, Castle Rock in background. X-10055 DPL WHC, used with permission.

In the early 1900s, the rise of aggregate quarrying and increased clay mining created demand for more mine-specific rail lines in Golden. Operators of the two quarries on North Table Mountain, the Tramway and Blake Quarries, constructed short, inclined trams to transport rock downhill to crushing facilities at railheads. The first tram (1907-1917) served the Tramway Quarry on the south side of North Table Mountain, with a crushing facility on the Denver Tramway Company (DTC) 83-line along North Golden Road/44th Street (Fig. 98). The operators of Blake Quarry built a tram[268] on the west side of North Table Mountain in 1909 to transport rock down to a crusher along the Brickyard Branch. Both trams were "gravity

Fig. 98. Rail car crossing North Golden Road (now 44th Street) on inclined tram from the Tramway Quarry on the south side of North Table Mountain between 1907 and 1917. View to west. From Robertson and Forrest (2010, p. 79), Colorado Railroad Museum Collection, used with permission.

---

[264] Newspaper reports are unclear whether track was laid. In addition, the mines in the Apex area (Chapter 10) reached high production volumes by the end of the 1890s, after the Financial Panic of 1893. Their actual start dates are uncertain.
[265] Litvak et al., 2020, has a good timeline of historic interurban (streetcar) rail lines in Colorado.
[266] They failed to electrify their railway due to lack of money. They continued to run dirty, noisy steam engines until a successor company electrified the system in 1909.
[267] Robertson and Forrest, 2010, p. 29.
[268] This was before the Glory Hole was dug in 1924.

operations." Operating with cables, the rail car carrying rock from the top was counter-balanced by an empty rail-car coming up from the bottom of the mountain. Single tracks divided into two tracks midway up the mountain to let the two cars pass eachother. The popularity of inclined trams in the early 1900s also led to building two funicular railways in Golden (Fig. 95): Castle Rock (1913-1919) and Lookout Mountain (1912-1916).[269] The funiculars operated similarly, but carried people, not rocks.

Fig. 99. Rail-bed embankments of the former Morrison Branch, east of the Golden Cemetery and Ulysses Street. Two weeks after taking this photograph, a development project removed the embankment. View to northeast in April 2021.

Built in 1910 by the Denver and Intermountain Company (D&IM), a subsidiary of the Denver Tramway Company, the last two mine-related branch lines built in the Golden area mostly transported clay. The Rubey Branch (Fig. 95) was built after years of complaining by downtown residents about dust and debris from wagon traffic carrying clay loads from the Rubey Mine on the south side of 12th Street. In addition, G.W. Parfet, Jr. had been trying to convince the D&IM to build a spur for several years to help mitigate those complaints.[270] In spite of chronic complaints, the 1912 construction of the CSM Experimental Plant, which needed rail access for its metallurgical research activities (Chapter 8), was the final motivation to build the Rubey Branch. In south Golden, the Morrison Branch was constructed in 1910 to serve the two large Apex (aka Queen City) brick-clay mines in south Golden. It was extended south in 1916 to serve the Parfet and the large Strainland fire-clay mines.

Ironically, this branch never reached the town of Morrison! The electrified Morrison Branch connected to the 84-line at Wyman station, near today's intersection of South Golden Road and Quaker Street (Fig. 95). Now buried under the Rooney Road Athletic Complex and the fill for I-70, remnants of the rail bed used to be visible until May 2021, along Ulysses Sreet, east of the Golden Cemetery (Fig. 99). Both branches lasted until 1953, when the D&IM/DTC pulled out the tracks.

The last mining-related rail line, the Brickyard Branch, was abandoned in 1963. Thus, the only remaining historic rail line is that of the original CCRR 1870 main-line from Denver to Golden that provides service for Coors Brewery today.

Mine Subsidence

After abandonment, underground clay and coal mine-workings tended to collapse over time, leaving holes, depressions, and generally unstable ground. Prior to 1980, mine subsidence in Golden was mostly on vacant range land. However, as Golden expanded its city limits by annexation, identifying areas of known and potential subsidence became mandatory. To this end, Jefferson County authorized a study of mine subsidence areas in 1977,[271] which turned out to be an essential basis for urban planning, especially as the study was conducted before urbanization.

---

[269] Warden, 2012.
[270] Parfet and Ryder, 1977.
[271] Amuedo et al., 1978.

Abandoned-mine subsidence in Golden ranges from obvious to subtle. The most obvious features involve trench scars from former fire-clay mines along the Dakota Hogbacks. Some of the trench scars still have overburden above uncollapsed stopes, creating hazards in the form of huge, hidden vertical drops. Today most of these are fenced off to prohibit public entry, even in designated open space areas (Fig. 100). All of the Dakota Hogback segments are now formal Jefferson County or City of Golden open space, but are closed to public access due to the mine hazard. In 2021, the State Mine Reclamation Board notified the City of Golden that open trenches left from the former Santa Fe fire-clay mine along Eagle Ridge will be filled in.

Fig. 100. In 2018, fences were installed around an abandoned vertical shaft at the former Tincup and South Golden fire-clay mines in south Golden after a near-tragic incident in 2017. View to south.

Highlighting the attractive-nuisance problem, in December 2017, two high school boys encountered near-disaster when one of them rappelled into an abandoned vertical stope along the Tincup segment of the Dakota Hogback (west of the Rooney Road Athletic Complex, Fig. 100). Using a clothesline for a rope, one of the boys got about 40 feet down when the rope broke, after which he fell 60 feet, breaking his leg. His friend called for help. The rescue operation took about about 3 hours and more than 20 personnel. Before the accident, neither boy thought the old shaft was particularly dangerous.

Subsidence features above abandoned and collapsed coal-mine shafts and lateral crosscuts (side tunnels) are more subtle problems. In north Golden in the area of today's New Star Way and Hogback Drive, subsidence pits (Fig. 101) were fresh in 1977 when the mine subsidence report was prepared. Some of the pits existed directly above the former vertical entrance shafts, whereas others were along lateral crosscuts that had collapsed. Now, the City of Golden regulates building construction specifically to account for issues caused by old coal mine workings.

Differential settlement is another legacy of mining in Golden, particularly on the west side of the Mines campus in the former Rubey Mine. Differential settlement refers to soft rocks compacting more than hard rocks under a weight. When a building straddles two natural materials of differing firmness, the weight of the building compacts the soft rock more than the hard rock. When that happens, a depression in the softer spot causes the foundation and walls to sink and crack. Human-made dirt (called "fill") that is filled into pits where mining has occurred can cause differential subsidence if a building or road is constructed across those filled locations.

Another part of the mine-subsidence legacy is how open-pit quarries were filled in prior to rules about mine reclamation; they made great places to dump trash. After the disastrous flood on Cherry Creek through Denver in 1965, flood debris was placed into the mined-out

Fig. 101. Fresh (circa 1977) subsidence pit (arrow) over shaft of abandoned Golden Star coal mine in north Golden on today's Brickyard Circle. Photo from Amuedo et al. (1978, p. VI-2a). View to northeast.

clay trenches in the former Rubey Mine.[272] The filled-in land was leveled off, making smooth-looking "cosmetically"[273] reclaimed ground. In the late 1960s after CSM acquired the Rubey Mine to extend its campus, married student housing was constructed across this ground, with poor consequences for building foundations and streets (Fig. 102). With the weight of the buildings, the old flood debris between the hard sandstone fins began to compact, causing parts of the buildings to settle and crack. All of the housing had to be demolished. The area is now an intramural athletic field and parking lot. Even now, swales form in the athletic fields and streets, which are then filled in with dirt and asphalt, respectively, to cosmetically smooth over the ground surface.

Fig. 102. Cracked and slumped sidewalk (black arrow) on the CSM campus circa 1977, caused by differential subsidence. Today's Stop 3 on the Weimer Geology Trail (white arrow) in background. Photo from Amuedo et al. (1978, p. VI-6a). View to north.

The Rooney Road Athletic Complex south of Colfax (U.S. 40) in south Golden occupies the now reclaimed and clay-capped Jefferson County garbage dump[274] that operated from 1958 to 1980 (Chapters 7 and 10). The dump used the site of the large Apex Clay Mine that operated from the late 1800s to about 1955. The mine was "reclaimed" by simply filling it with garbage from Jeffco residents. Buried garbage rots over time, forming methane that, if not vented, tends to explode. When the Athletic Complex was being built, welding was not permitted onsite due to that danger of explosion.[275] Vents now surround the landfill, releasing methane to the atmosphere.

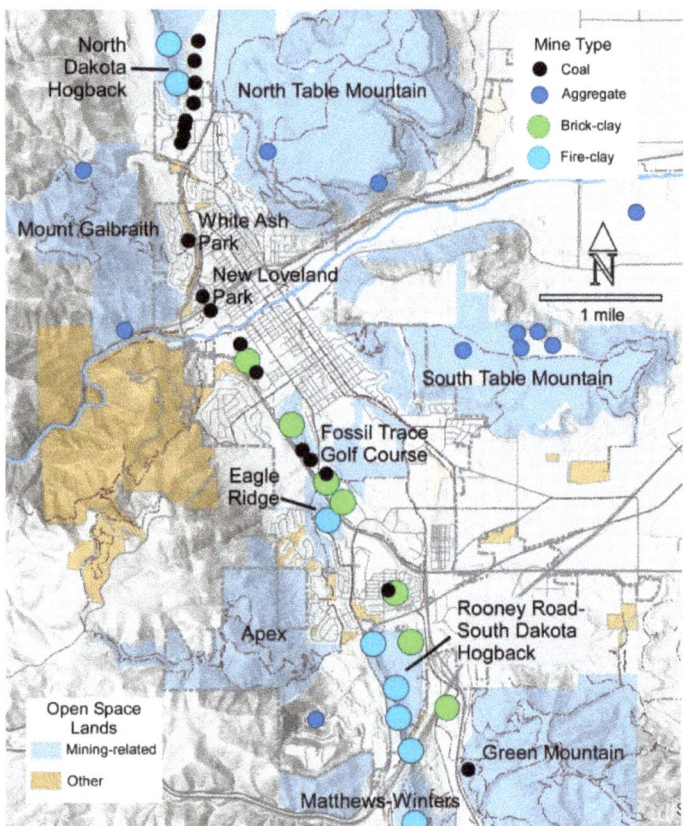

Mining History and Open Space

Did you know that 90% of the parks and open space in and around Golden has historic connections to mining (Fig. 103)? For example, New Loveland and White Ash Parks are the former sites of coal mines with the same names. The two parks were created when Coors property in the Canyon Point area was developed in the 1990s. They form

Fig. 103. Open space lands with historic connections to mining around Golden. Boundaries compiled from City of Golden GIS data.

---

[272] Ibid., p. VI-6, and Parfet, 2007.
[273] "Cosmetic placement" is a euphemism for just dumping dirt and trash in a hole without engineered compaction.
[274] Our modern sensibilities dictate that garbage dumps be called "sanitary landfills." Regardless of name, they still generate methane gas from rotting garbage, and the gas from the landfill must be vented to prevent explosions.
[275] Carder, 2007, describes the special construction requirements.

buffers around abandoned coal-workings to guard against mine subsidence. Several of the Parfet family-operated clay mines became open space of various types. Fossil Trace Golf Course is the site of the former Rockwell Mine (Fig. 104). Instead of filling in the old quarry, the land was re-shaped and landscaped. The sandstone "fins" in the other fairways of the "back-nine" holes were left behind to form scenic golf hazards. Other parts of the former mine area were left to preserve the dinosaur and plant trace fossils, now part of Triceratops Trail (Chapter 4).

Fig. 104. 1977 view of the then-active Rockwell Mine and what is now the 15th Fairway at Fossil Trace Golf Course. Sandstone cliffs are over 40 feet high. Photo from Amuedo et al. (1978, p. VI-7a). View to northeast.

The Rooney Road Athletic Complex is the site of the former Apex Mine and the reclaimed Jeffco garbage dump. The Dakota Hogback, west of the Athletic Complex, is all part of Jefferson County Open Space, as is the former Santa Fe Mine along Eagle Ridge. In north Golden, part of the former Golden Pressed Brick Mine within the Golden city limits forms the North Hogback Park and Open Space. Apex Park in south Golden is the site of an historic wagon road leading to the old mine workings around Idaho Springs. The west side of Green Mountain contains abandoned coal and clay mines, as well as the currently (2021) active North and South Chieftain Mines.

Some of the largest open space parcels are those with long aggregate-quarrying histories or sites proposed for quarrying. Post-1970 histories of these aggregate quarries follow the tale of changing residents' attitudes regarding mining and the environment that began with the environmental movement of the late 1960s in the United States.

Collisions

Since 1970, the history of crushed-rock aggregate quarrying in Golden involved increasing resident opposition to mining operations. Two issues, operating at the same time, collided to create controversy that continues today: increasing population growth, resulting in urban sprawl; and increasing demand for mining products, especially aggregate.

After World War II, the population in Golden, along with the rest of the U.S., grew rapidly (Fig. 105). Golden expanded its city boundaries, encompassing more land for housing and business developments. After 1970, within and around Golden, urban developments covered more and more formerly vacant lands. As homes were built near existing quarries, residents objected to dust and air pollution, truck traffic and noise, and severely altered scenic vistas. In 1971, a

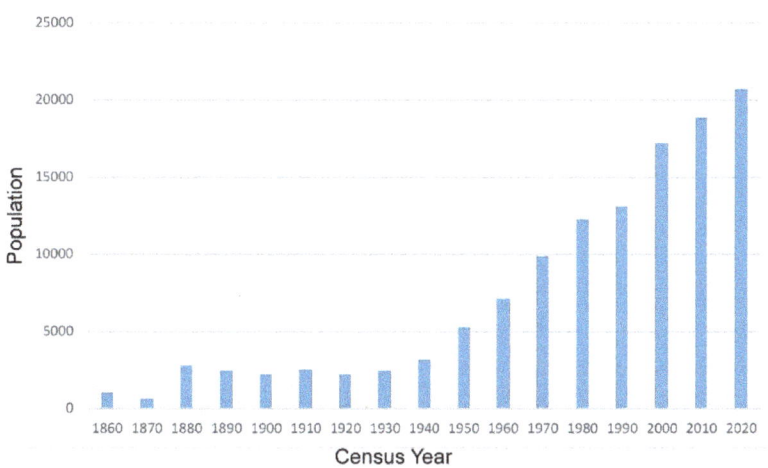

Fig. 105. Population of Golden 1860-2020 from U.S. Census data. Note the rate increase beginning in 1950.

citizen-led group, PLAN Jeffco,[276] promoted preservation of open space in Jefferson County. In 1972, a Jefferson County ballot initiative authored by PLAN Jeffco and supported by the League of Women Voters, was approved by the voters. It created and <u>funded</u> Jefferson County Open Space, with the mission to preserve open space and parkland; protect park and natural resouces; and provide healthy, nature-based experiences within the County.

At the same time, increased construction of housing, commercial/industrial buildings, and the roads to serve them, led to increased demand for crushed rock (aggregate). Between 1970 and 1976, proposals to expand aggregate quarrying on North and South Table Mountains met with unwavering local resistance that stopped those proposals. Those in the aggregate-quarrying business realized that quarries and quarry sites were disappearing due to advancing urbanization. By 1977, a task force for the Jefferson County Planning Department created a Mineral Extraction Policy Plan designed to control the impacts of urbanization and help preserve aggregate resources. However, the plan did not address <u>how</u> to protect aggregate resources from progressive urbanization.[277]

Between 1976 and 1997, local resistance increased as more quarries were proposed on North and South Table Mountains and along the Front Range. As a result, since 1977, no applications for <u>new</u> quarries have been approved in Jefferson County. In addition, most proposals to re-open former quarries have been denied. Understanding that "vacant" land without the proper zoning/ownership was subject to continuing development proposals, Jefferson County Open Space became the organization that purchased proposed quarry areas. Using the funds generated by the 1972 initiative, acquisition gave the vacant lands a legal status as dedicated open space.

The list of quarry-related Jeffco Open Space acquisitions around Golden is short, but consists of the largest open-space parcels shown in Figure 103, including Mount Galbraith and North and South Table Mountain Open Space Parks. The land for Mount Galbraith Open Space Park was purchased by Jeffco Open Space after a contentious proposal to open a new quarry between Clear Creek and Golden Gate Canyons. North Table Mountain Open Space was acquired in a land swap with LaFarge North America for more land to operate the Spec-Agg Quarry in south Golden. South Table Mountain Open Space was acquired from Coors after defeating proposals to quarry and then develop the mountain top for a business park.

Today, operators try to expand pre-existing quarries. Such proposals usually involve a lot of controversy. For example, land that is now part of the Matthews-Winters open space near the Spec-Agg Quarry, is under consideration for mine expansion (in 2021). The operators propose a land-swap involving current open space south of the quarry and north of I-70 (Chapter 11). The Spec-Agg Quarry itself is slated to become Jefferson County Open Space when operations cease.

Has the need for aggregate decreased? No. Have the issues around opening or expanding quarries changed? No. The few operating aggregate quarries now within the urban area, like the Spec-Agg quarry, have become extremely valuable. Aggregate now comes from quarries that have been operating for the last 30+ years and/or are located in relatively un-urbanized areas. Usually these are farther away from the urban area, requiring longer truck routes. The situation is one of a paradox: needing quarry products for urban growth <u>and</u> maintenance while not desiring quarry operations in or near urban areas.

Future Musings

Predicting the next 25 years, let alone the next 50, is a task for fortune tellers. The population-growth trend suggests that land will continue to be covered by urbanization, which will cover and/or alter geologic, paleontologic, and historical resources. What is more, not understanding history, either deep time or human time, limits our understanding of context going forward. Certainly, understanding

---

[276] In 1971, PLAN Jeffco began in the living room of Jeffco resident, Carol Karlin after a study done by the League of Women Voters showed a need to conserve open spaces in Colorado. Funding for open space acquisitions comes from a ½% sales tax on items purchased in the County. Deserving of its own book, some of the history is at https://www.jeffco.us/1571/About-Us

[277] Langer and Tucker, 2003, p. 5.

our water resources within the context of our semi-arid climate is also important to our future water use, especially as drought has become more common since 1950.

Golden's constant geologic sentinels, North and South Table Mountains and the Front Range escarpment, keep Golden geographically unique within the Denver Metropolitan area. Yes, geology has shaped the Golden landscape, and since 1859 Euro-Americans have complicated the tapestry of that landscape. We hope that this book illuminates historic context at all time scales, and engages you, the reader, to go outside, and see for yourself. Understand the landscape around you and beneath your feet, and learn the stories on how the landscape came to be.

Acknowledgments

Some say "it takes a village" to write a book. Nothing could be truer in this case, especially as we researched through the 2020-2021 pandemic, when physical access to libraries, museums, and other archival organizations was prohibited. We greatly appreciate reviews by Larry Anderson, Dr. Mark Longman, Bob Bruner, Dr. Dennis Gertenbach, Dr. Holly Huyck, Dr. Bob Raynolds, Kermit Shields, and Jane Estes-Jackson who helped us stay on track technically and with our writing style for a general audience.

Periodic discussions with Dr. Robert (Bob) J. Weimer[278] (Emeritus Professor at CSM), Dr. Bob Raynolds (Denver Museum of Nature and Science and Friends of Dinosaur Ridge), Ned Sterne (Geologic Consultant), Ann Norton (Stewards of Golden Open Space), Barb Warden (GoldenToday.com and one of our historical muses), Frank Blaha (Environmental Engineer), Dr. Bruce Trudgill (Geology Professor at CSM), Dr. Holly Huyck (Environmental Geoscientist, for Chapter 7), Kermit Shields (Friends of Dinosaur Ridge), Lisa Dunn (Head of Special Collections, Arthur Lakes Library), Bill Robie and Bill Brown (Colorado Railroad Museum), Dan Abbott (author and railroad fan), Bill Litz (Golden Landmarks Association), Jack and Kenny Brunel, Peter J. Coors and son, Peter H. Coors, W. G. (Chip) Parfet, Jr., and Richard Gardner (Golden Historian) helped us with a myriad of details and questions. In addition, field trips and talks were invaluable for the commentary and questions from participants from all walks of life. Such trips and talks included in-person field trips for the Colorado Scientific Society and "The Geologic Study Group," a virtual field trip for the Rocky Mountain Association of Geologists in Fall 2020, and in-person community-based walks with the Stewards of Golden Open Space. We also gave virtual talks for Golden United in April-May 2020 and blogged for Barb Warden's "Golden History Moments" column in GoldenToday.com.

Online research was essential to our work. It is amazing what is available online, and that was even more critical during the 2020-2021 pandemic. We acknowledge the online collections of Colorado Historic Newspapers, Internet Archive, Denver Public Library Western History Collection, USGS photographic archive, USGS topographic and geologic maps, Golden History Museum, Historically Jeffco, Foothills Geneaological Society, Arthur Lakes Library (CSM) Colorado Digitization Project of the Russell L. and Lyn Wood Mining History Archive, and ESRI Online for map imagery. We also had helpful research assistance from Chris Thiry (Map Librarian) at Arthur Lakes Library, Mark Dodge (Curator) at the Golden History Museum, Stephanie Gilmore (Librarian) and Larry Anderson (volunteer) at the Colorado Railroad Museum, Melanie Keerins (Corporate Archivist) at CoorsTek, Kim Soulliere (GIS Coordinator) at the City of Golden, Kellen Cutsforth at the Denver Public Library, Jori Johnson at the Stephen H. Hart Research Center at History Colorado, and Rhonda Frazer (County Archivist) at Jefferson County, Colorado.

We also thank the following for permission to access geologically important outcrops in and around the City of Golden: Martin Marietta Aggregates to enter the former Heritage Square property, and the City of Golden to enter the city shops off Catamount Street. Otherwise, most of the outcrops shown as figures in the book are accessible to the public without permission at the time of writing. Organizations and individuals granting permissions to use photographs are cited in the caption of each photograph.

---

[278] Dr. Weimer passed away in August, 2021, just before we finished the book. He was highly regarded as a gentleman and a scholar, and he touched many people's lives. We will sorely miss him.

Glossary

Accuracy: refers to how close a measurement is to the true or accepted value of something. See also Precision.

Activity (as in earthquake activity): The recurrence of earthquake along a fault or group of faults. Frequent activity would have a lot of earthquakes in historic times. Low activity would mean that earthquakes had not occurred much over the last 12,000 years.

Adit: a horizontal entrance tunnel leading into a mine.

Alkaline: in chemistry, refers to a basic solution, with a pH more than 7: the opposite of acid.

Alluvial Fan: a triangle, or cone, shaped accumulation of sediment, such as sand and gravel, usually at the mouth of steep canyons, formed when water rushes out of the canyon and slows down, dropping out the sand and gravel in a cone.

Alluvium: a deposit of clay, silt, sand, and gravel left by flowing streams across the land surface, mainly in a river/stream valley

Andesitic: a volcanic rock composition tending toward moderate amounts of silica, like the rock andesite, which is between basalt (low silica) and rhyolite (high silica). The rock type is named after the Andes Mountain Ranges of South America.

Angular unconformity: a type of unconformity (see also "unconformity") in which younger flat rock layers were deposited over older tilted, eroded rock layers

Anticline: a ridge- or arch-shaped fold in which the layers are bent downward from the top.

Aquifer: a body of rock that contains and can transport groundwater through tiny holes or cracks in the rock.

Arthropod: an invertebrate animal like an insect or spider. The word comes from the Latin, meaning "those with jointed feet."

Back-thrust: A thrust fault in which the displacement is in an opposite direction to that of the main thrust fault. See also thrust fault.

Basalt: a dark fine-grained volcanic rock, may form lava on the ground surface, in which case it is called a basalt flow or a basaltic lava flow.

Basin: A large low-lying area that collects sediment. Can either be a modern topographic feature, like the Pacific Ocean Basin, or a valley like the Denver Basin. It can also be a geologic structural feature that is down-folded and contains thick layers of rock, also like the Denver Basin which in the subsurface covers all eastern Colorado.

Bed, Bedding: in geology, a layer of sedimentary rock that is distinctly separate from what is above and below it. Bedding simply refers to several beds that form a series of layers.

Bedrock: in geology, the solid rock that is buried beneath soil and other loose, unconsolidated (not rock yet) material.

Brackish: refers to water that is slightly salty. Common where seawater mixes with freshwater, like in an estuary.

Brittle: here used to describe material that is hard and breaks or shatters easily. Rocks can be both brittle and plastic!

Bullion: gold or silver in bulk before turning it into coinage or jewelry.

Caprock: A harder or more resistant rock type overlying a weaker or less resistant rock type. An analogy would be the outer crust on a cake that is a bit harder than the underlying layer.

Chalk: a sedimentary rock composed of microscopic organisms (usually a type of algae) composed of calcium carbonate. The rock forms the major component of chalk used to write on blackboards!

Chert: A hard, dark sedimentary rock composed of micro-crystalline quartz (silica). Also called flint.

Clastic, clasts: denotes a sedimentary rock composed of grains, or pieces, of older rocks. The grains are also called clasts. The grains can be any size ranging from ultra-tiny to more than ten feet in diameter.

Colluvium: a general name for loose, unconsolidated sediments deposited by a variety of processes.

Compressive (compression) tectonic forces: large Earth forces that squeeze over huge areas, both vertically and horizontally.

Contact: in geology, a boundary separating one rock body from another.

Crevasse splay: a sedimentary deposit formed when a river breaks through its bank and sediment spills out into the floodplain. Where the river breaks through is the "crevasse" which is French for "breach." The splay is the mounded body of sediment deposited when the sediment spreads out from the breach.

Crust, Crustal: the outermost solid layer of the Earth. It varies in thickness from 5 to 40 miles thick, depending on if it is in the ocean (thin) or on continents (thick). It is composed of a great variety of rock types. Crustal refers to rocks within the Crust.

Crystal lattice: the microscopic, highly ordered arrangements of atoms and molecules that make up a solid mineral.

Deep time: the multi-million-year time frame within which the solar system has existed.

De-glaciation: the uncovering of land and water that was previously covered by glacial ice.

Deposition: the process of laying down sediment carried by wind, water, or ice. Sediment can be carried as pebbles, sand, and mud, or as salts dissolved in water.

Depositional environment: a specific type of place in which sediments are deposited, such as a stream channel, a floodplain, a lake, or ocean bottom.

Dike (in geology): a sheet of molten igneous rock that is intruded into nearby solid rock under great pressure, and cuts across that rock.

Dip/dipping: Simply the tilt of beds (see bed) in sedimentary and/or extrusive igneous rocks, like lava flows.

Discharge, dischargers (of water): in hydrology (study of water) it is the volume of water that flows through a given area at a given velocity. Given by Darcy's Law: Discharge (ft^3/second = Area (ft^2) x Velocity (ft/sec). Dischargers are people or organizations who discharge water.

Dissected: cut up, usually from stream erosion

Dredging: in mining, the underwater excavation of a placer deposit by floating equipment.

Dysoxic: having a very low oxygen concentration, which inhibits living organisms.

Epicenter: as in earthquake epicenter, the point on the Earth's surface vertically above the actual first point of rock breakage, where an earthquake begins below the ground. See also focus.

Erosion, erosive: in geology, the process by which Earth materials are worn away and transported by wind and/or water. These processes are called "erosive."

Estuary: a partly enclosed coastal water body of brackish (slightly salty) water with one or more streams flowing into it, and a free connection to the open sea.

Exhumed: un-buried.

Extrusive: the mode of volcanic rock formation in which molten rock (magma, see below) flows out on the surface (extrudes) or explodes violently into the atmosphere and falls as volcanic ash.

Fault (in rocks) and Fault plane: the crack, or series of cracks, between two rock-bodies that has allowed the two blocks to move (see offset) by eachother. See also the Geology Primer.

Fault line (same as fault trace): the place where the edge of a subsurface fault intersects the ground surface.

Feldspar: a group of minerals composed of alumina and silica with the elements calcium, sodium, and/or potassium in the crystal lattice. It is the most abundant mineral group on Earth.

Fire-Clay: A large range of refractory clays used in making ceramics, especially heat resistant ceramics. Such clays are also composed of aluminum silicates, and tend to have more aluminum in the crystal lattice.

Flume: a human-made channel, made of steel or wood, that carries water through it.

Focus: as in earthquake focus, the initial point of rock breakage during an earthquake below the ground. Also called the hypocenter.

Fold: a bend or curve in rock layers, or strata, of any type. Size can vary from a few inches to miles. See also the Geology Primer.

Fractures (in rocks): a crack dividing a rock into two sections. The sections have simply split apart with no movement, or offset (see below), between them. They naturally occur when long-buried rock is brought to the surface and the overburden forces are released. They can also be made by injecting fluid under very high pressure into the subsurface, called "fracking," breaking the rock. Also called "joints."

Geomorphologist: An Earth Scientist who studies landforms at the Earth's surface and the processes that made them.

Glacier: A large, multi-year to thousands of years, buildup of crystalline ice on the Earth's surface that moves down slope under the influence of its own weight and gravity. Can form continent-scale sheets of ice.

Gneiss: a metamorphic rock that has been subjected to very high temperatures and pressures, more so than its relative, schist (see below). It has a characteristic banded look, in which the bands are alternating dark and light layers composed of different minerals.

Gradient: as in river and stream gradient, the slope of the stream bed. Measured in by the difference in elevation between two points along the stream.

Granite: a light-colored intrusive (see below) igneous rock with crystals large enough to be seen by the naked eye. It forms from slow cooling of molten rock below the Earth's surface, commonly in large masses.

Gravity system: any human-made system that relies on natural gravitational Earth forces for its motion. In water supply, it a system of pipes or ditches in which water flows downhill under gravity and without the use of pumps or other external energy to deliver that water. Gravity-tram trains similarly rely only on gravity to move.

Headgate: in water supply, especially irrigation, a structure to control the flow of water into a channel or ditch.

Hematite: a reddish-black mineral consisting of ferric (iron) oxide. It becomes brick-red when it oxidizes, aka "rusts."

Hydraulic mining: a form of mining, commonly for gold, that uses high pressure jets of water to dislodge rock material, commonly unconsolidated sediment such as alluvium. The resulting stream of water and sediment is directed through sluiceboxes (see below) to remove the gold from the sediment-water mixture.

Igneous: rock that has solidified from lava or magma. Word comes from "fire" or "fiery."

Interbedded: lying between and commonly alternating beds, or layers, of different sedimentary rocks.

Interglacial: times when ice sheets (glaciers) are absent: they have largely melted away, resulting in deglaciation.

Intermittent: a stream that stops flowing during the dry season, also called "seasonal."

Intrude(d), Intrusion, Intrusive: the mode of igneous rock formation in which molten, underground rock (magma) pokes into, or intrudes, other rocks in the Earth's crust below the ground surface, then solidifies while cooling. Intrusion is the resulting rock body, like a dike or a sill. See Geology Primer.

Isotope: An element, such as argon, with the same number of protons, but with differing amounts of neutrons. Argon, for example, has 26 known isotopes ranging from ^{29}AR to ^{54}AR, only three of which are stable on Earth. The most abundant argon isotope is ^{40}Ar.

Kaolinite: A soft, earthy white or gray clay mineral with the chemical composition of $Al_2Si_2O_5(OH)_4$. A highly desirable clay mineral for making fine China and ceramics. The word comes from the Chinese word "gaoling."

Lava: hot molten or semi-molten rock erupted from a volcano or fissure that has cooled and become solid.

Magma: hot, fluid to semi-fluid rock occurring below the Earth's surface. It is the starting point for all igneous rocks.

Magnitude: as in earthquake magnitude, the measure of an earthquake's size and energy where the earthquake begins.

Mantle: the part of the Earth between the crust (see above) and the core. It is about 1800 miles thick and makes up nearly 80% of the Earth's volume. It is made up of magma (see above) and solid rock.

Mountain-building: as in mountain-building episode: the process in which a section of the Earth's crust is folded, faulted, sometimes metamorphosed, and generally deformed and uplifted to build a mountain range. Same as orogeny. See also the Geology Primer.

Metamorphic/Metamorphosed: relating to a rock that has changed by heat and/or pressure without wholesale melting. See also the Geology Primer.

Moraine: an accumulation of rock debris carried and deposited by a glacier. Moraines are common where a glacier ends down valley, creating a series of looping ridges around the glacier edge. They also mark where glaciers advance the farthest.

Non-refractory: as in clay, a clay that easily changes shape or melts at normal kiln temperatures. The opposite of refractory clays.

Nutrient: a chemical substance that provides nourishment for growth. In this book, it refers to fertilizer compounds that nourish algal growth that pollutes water bodies.

Ocean basin: a vast submarine region. Today one would think of the deep, vast floors of Pacific and Atlantic Oceans as oceanic basins. In the geologic past, ocean basins existed where land is today, such as the Western Interior Seaway discussed in this book.

Oceanic plate: a thick mass of igneous rock that underlies the ocean floor.

Offset (as in offset along a fault): the dislocation, or displacement, along a fault plane. Horizontal displacement is measured on the land surface between points on either side of a fault. Vertical offset is measured in the subsurface (underground) between two known points.

Orogeny: the process in which a section of the Earth's crust is folded, faulted, sometimes metamorphosed, and generally deformed to build a mountain range. See also the Geology Primer.

Outwash (as in glacial outwash): Sand and gravel deposited from running water resulting from the melting ice of a glacier and laid down in bedded layers.

Overturned: as used in this book, refers to the condition in which sedimentary beds are rotated into positions in which the bed becomes upside down: caused by structural deformation.

Paleontology: the scientific study of life of the geologic past that involves the analysis of plant and animal fossils preserved in the rocks. Paleontologists are the people who study the fossils.

Pegmatite: an igneous rock formed underground with interlocking crystals more the 1 inch in size. Most have a similar composition to granite, and they intruded the Earth's crust as a dike, cutting across pre-existing rock, at the last stages of magma cooling. Very slow cooling allowed for forming huge crystals. Gem-quality minerals like beryl and tourmaline are found in some pegmatites.

Perennial: A river or stream that flows continuously in parts or all its stream bed year-round during years of normal rainfall or snowmelt.

Placer: a deposit of sand or gravel in a river or stream bed containing particles of valuable minerals, such as gold.

Plastic: as in structural deformation: The permanent distortion that occurs when a material, like rock in this case, is subjected to bending, tension or compressive forces that make it deform. Below ground, rocks can behave plastically, especially when heated.

Plate tectonics: a theory explaining the structure of the Earth's crust as resulting from the interaction of plates composed of the rigid rocky outer layer of the Earth (the crust plus the outermost layer of the mantle), which move slowly over the underlying mantle.

Pores: tiny holes, or voids, between sedimentary grains or along joints within a rock, which can contain air, water, oil, natural gas, or other fluids. The percentage of the pore spaces versus the actual rock is the porosity of the rock.

Portal: the starting point of a shaft or tunnel.

Precipitation (chemical): in chemistry, the process by which a substance is separated out of solution as a solid via a chemical reaction.

Precision: refers to how close measurements of the same item are to eachother. See also accuracy.

Quartz: a mineral composed only of silicon and oxygen, SiO^2, that forms the building block of a large group of minerals known as "silicates." Quartz is highly resistant to chemical breakdown, or weathering. As such it is a common component of all rocks.

Quartzite: an extremely compact, hard, granular rock consisting of quartz. Commonly formed by metamorphism of sandstone.

Refining: in petroleum, the process of reducing the impurities in petroleum by heating and then cooling to let different impurities and oil densities drop out.

Refractory: as in clay, a clay the does not melt at normal kiln temperatures: that is, it withstands very high temperatures and keeps its shape, which is very desirable for making ceramics.

Retorting: in mining, part of the process of extracting liquids, such as petroleum, from a rock by heating to a high temperature, where the petroleum liquifies and comes out of the rock.

Reverse fault: a fault in which the upper block above a fault plane moves up and over the lower block. The fault plane usually dips (is tilted) to more than 45°.

Rhodochrosite: a mineral consisting of manganese carbonate, typically as red, pink, brown, or gray crystals (rhombohedral shape). Rhodochrosite is a striking rose-red color.

Rift: in geology, a linear zone where the Earth's crust is being pulled apart, forming a down-dropped basin between two parallel fault zones.

Riprap: a foundation or sustaining wall of stones or chunks of concrete placed together without order.

Rock formation: in geology, a group of rocks with some type of similar characteristics, like rock type, that can be mapped at a scale of about 1:24,000, or 1-inch equals about half a mile. So, defining the group is a bit arbitrary, which has created a lot of confusion over the decades.

Schist: a metamorphic rock that has medium to large, flat, sheet-like grains that are roughly parallel to eachother. It forms under lower temperature and pressure than gneiss.

Seam(s): A mining term for a sedimentary bed, which is simply a layer of limestone, sandstone, or coal that begins and ends within a mining area. The longer the "seam," the more productive the mine.

Sediment, sedimentary: solid particles that move and/or settle to a new location: can consist of rocks, minerals, and/or plant and animal remains. Sedimentary: a rock that has formed from sediment deposited by water, air, biological buildup, or chemical precipitation like salt. See also Geology Primer.

Septic system: an underground wastewater treatment structure used in rural areas without centralized sewer systems. Consists of a tank that collects wastewater, treats it, and then disperses it to a series of underground pipes that let the treated wastewater seep out into the ground in a person's yard.

Seismic/Seismograph: pertains to earthquakes or other vibrations of the Earth and its crust. A seismograph is an instrument that measures the vibrations.

Senior (water) rights: the first person to use water on a stream or river acquires the right to its future use against later (junior) users. In Colorado, a senior user may place a call on a river or stream because the senior used is not getting the water it is entitled to. Those with junior rights are shut off until the senior right is satisfied.

Siphon: in terms of a pipeline or tunnel, it is a tube used to convey water from the same elevation across a valley or gulley, letting gravity do the work. It is another type of gravity system.

Slickensides: the grooves made a long a fault plane, where the frictional melting and grinding of rock moving on either side of the fault, creates scratches and/or a polished fault surface.

Sluicebox: in mining, a long sloping trough with slats and/or grooves on the bottom, into which heavier gold drops out of moving water, and gravel and sand wash out of the box.

Smectite: a group of platy clay minerals, usually less than 8 microns in size, that can soak up a lot of water molecules within the crystal lattice, which in turn makes them swell when wetted.

Soil horizon: a layer at the surface of the Earth in which you may garden. Formed by natural, in-place break up of material at the Earth's surface, usually formed over a period of landscape stability. As these have formed throughout Earth's history, they can be preserved in the rock record, where they are commonly called "paleosols," or "old soils."

Stope (rhymes with "mope"): a steeply inclined, subvertical shaft in a mine, used to remove ore that has been made accessible by lateral, horizontal tunnels.

Stoping: (rhymes with "moping") a gravity-based mining method in which the ore is blasted with dynamite and then allowed to cave-in (in a controlled manner, we hope) to the floor of the mine and then removed.

Strain gauge: basically, an instrument to measure tiny changes in shape, such as when rocks start to cave-in or fall-down in mines.

Strata: the plural of stratum, means the layers of sedimentary rock or soil, or igneous rock that were originally deposited at the Earth's surface, with internal similar characteristics that distinguish it from other layers.

Stratigraphy: a branch of geology concerned with the study of rock layers (strata) and layering (stratification). "Stratigraphic" means pertaining to stratigraphy.

Structural deformation: twisting, bending, and breaking at the size of a mountain front, like the Front Range. Usually refers to individual geologic structures, like folds and faults. Also, refers to the squashing and breaking of the size or shape of a rock, caused by Earth forces

Structural geology: a branch of geology concerned with the study of 3-dimensional distribution of deformed rock units and layers and how they formed.

Structures (geological): folds, faults, and fractures that form within rocks by deformation and are associated with mountain-building.

Subducting (oceanic) plate: a tectonic plate formed of ocean crust that dives underneath another tectonic plate composed of oceanic and/or continental crust. Subduction is the sideways and downward movement of the edge of a tectonic plate under another plate. The underlying plate pulls itself down into the mantle by gravity.

Sulfide ore: Rocks containing metals that occur in chemical association with the element sulfur.

Superimposed: generally, means something is laid over top of something else, and is quite different from what is beneath it.

Surficial deposit(s): Thin, unconsolidated "dirt-like" sediment that overlies pre-Quaternary bedrock. Term is used by Quaternary geologists to generally denote non-bedrock sediments.

Syncline: a trough-shaped fold in which the layers tilt down into the trough.

Tailings: The ground-up waste rock left over at the surface from mining operations. At the time of mining, they have no remaining economic value.

Tectonic: relating to the structure of the Earth's crust and the continent-sized processes that take place within the crust. Also refers to the forces or conditions within the Earth that cause crustal movements.

Thrust fault: a reverse fault with a moderate to low dip (tilt) of the fault plane, usually less than 45°.

Tongue (of sandstone): in sedimentary geology, particularly in stratigraphy, used to describe a mile or so long, sandstone layer or series of layers, that pinches out (terminates) into a group of shale layers.

Triangle Zone: a triangular-shaped structural deformation zone below the ground surface that is bounded by thrust faults dipping (tilting) in opposite directions. Generally, the earlier main thrust faults are cut off by the later back-thrusts.

Unconformity: The line, or contact, separating two rock masses of different ages. The contact represents an interval of rock that is missing either by erosion or never having been deposited. Thus, a contact represents an amount of geologic time: short, as in less than a million years, or huge, like billions of years.

Unconsolidated (deposits): formed by the initial result of natural weathering processes that break down large rocks into smaller ones. The resulting loose material is no cemented by natural processes, either.

Vein: in geology, a distinct sheet-like body of crystallized rock material intruded into a large mass of rock. Miners in Golden commonly use the term to describe coal and clay seams, which is technically incorrect, but so what?

Viscous: having a thick, sticky consistency. For example, syrup is more viscous than water.

Volcanic ash: the very small particles of rock and glass thrown out of an erupting volcano that eventually fall to the Earth's surface. Being very light, they can travel hundreds of miles from the volcano, finally being deposited on the land surface or on the ocean bottom as thin layers.

Vug: in geology, a natural hole in an extrusive volcanic rock, like a lava flow, formed by gas bubbles. The hole may be lined with mineral crystals.

Water cycle: The movement of water, including water vapor as well as liquid and ice, through and across the atmosphere, ground surface, and underground. The complexity is explained at this website, among many others: https://www.usgs.gov/special-topic/water-science-school

Watershed: A land area that channels rainfall and snowmelt to creeks, streams, and rivers, eventually to outflow points such as other streams and rivers, as well as reservoirs and ultimately to the ocean

Weathering: In geology, the in-place breakdown of rocks at the Earth's surface by the action of rainwater, temperature extremes (freeze-thaw), and biological activity. The three types of weathering are physical, chemical, and biological.

Zeolite: a family of minerals made of aluminum and silica arranged in a very open crystal lattice, or arrangement of atoms making up the mineral framework The open lattice allows other elements to come and go easily, making the mineral useful for industrial uses like purifying drinking water.

Geology Primer

Geology, like any field, has a lot of words with distinct meanings. So first, we need to lay out some basic terms and concepts to gain a shared language of geology. In the footnotes, you can find website hyperlinks to resources to learn more about geology. As in the main body of this book, words highlighted in blue are in, but not hyperlinked to, the Glossary.

Three Types of Rocks

Essentially, there are three types of rocks,[279] and we have all of them in Golden. The following discussion relies on excellent definitions from the U.S. Geological Survey.[280]

- Igneous: formed from the cooling of molten rock.
- Metamorphic: formed through the solid-state (not melted) change of igneous, sedimentary, and even other metamorphic rocks ("metamorphosis" means "changing form").
- Sedimentary: formed by cementing together particles, or sediment. Three types of sedimentary rocks are clastic, biologic, and chemical, discussed below.

Igneous Rocks

Igneous rocks form when hot, molten rock crystallizes and solidifies. Originating deep in the Earth's the mantle, over 100 miles below the crust,[281] the molten rock, or magma, rises toward the surface. Depending on where the magma solidifies, it is either intrusive or extrusive. If it cools and crystallizes deep underground, the rock is termed intrusive igneous rock, of which granite is a common example. If it solidifies at the surface, it is extrusive. Also known as volcanic rocks, extrusive rocks form from erupting volcanoes and/or cracks/vents. Extrusive flowing out of the volcano, crack, or vent, the forms lava flows along the ground surface. If thrown into the air, ejected lava forms volcanic "bombs" and volcanic ash. When the lava cools and solidifies, it becomes a rock. A common example of cooled lava is basalt, such as the rock atop North and South Table Mountains.

Metamorphic Rocks

Metamorphic rocks, like sedimentary rocks, started out as some other type of rock, but, as you might guess, they have undergone a substantial change, or metamorphosis, from the original rock. Originally, they could have been sedimentary, igneous, or even another type of metamorphic rock. As they undergo this transformation deep underground, the rocks become denser and more compact. The presence of metamorphic rocks commonly indicates a mountain-building period (see Rock Cycle below).

Metamorphism does not melt the rocks. Instead, hot fluids and/or extreme pressures rearrange the mineral components without melting. Metamorphic rocks commonly show squashed, smeared, and folded textures. In the mountains on the far western side of Golden, metamorphic rocks are represented by gneiss (pronounced "nice"), a rock where the minerals separate into bands during metamorphism. The State Rock of Colorado (yes, there is a State Rock) is a metamorphic rock called marble or more specifically Yule Marble from Marble, Colorado.[282] It was metamorphosed from limestone, a sedimentary rock.

---

[279] Rocks are composed of minerals which are naturally occurring crystalline solids formed by elements, such as quartz which is silicon dioxide (SiO_2). Add more elements and you get a different mineral and crystal forms, like biotite mica which is sheets of $K(Mg, Fe)_3AlSi_3O_{10}(F,OH)_2$. Pure carbon in crystalline form is called diamond. https://en.wikipedia.org/wiki/Mineral has more explanation.

[280] For example, this hyperlink talks about igneous rocks: https://www.usgs.gov/faqs/what-are-igneous-rocks?

[281] Basically, the Earth has a crust, mantle, outer core, and inner core. The crust is solid, the mantle is a highly viscous solid, and the inner core is a solid. Only the outer core is a liquid. See https://en.wikipedia.org/wiki/Structure_of_the_Earth for more on the structure of the Earth interior.

[282] Marble from Marble, CO, was used to build the Lincoln Memorial in Washington D.C. For more on Colorado's official rocks and minerals, as well as the geology of the whole state, see the award-winning book, "Messages in Stone" by Vince Matthews (2003), former director of the Colorado Geological Survey, and his colleagues.

Sediment, the component of sedimentary rock, is formed in three ways. First, from particles (pieces) of pre-existing rocks (igneous, metamorphic and pre-existing sedimentary rocks), called clastic sediment. Second, from the remains of once-living organisms, called biologic sediment. Third, from chemical reactions in water that form chemical sediment. Particles of clastic sediment form by loosening from mechanical or chemical break down (weathering) of pre-existing solid rocks of any type. In the case of biologic sediment, material from dead organisms, also called fossils[283] (usually shells or plants), basically piles up, or accumulates, where it died. Chemical reactions may form crystals of chemical sediment by precipitation, such as salt and calcium carbonate. Any type of sediment may be moved (transported) from its original site to be deposited at another location by wind or water where it finally accumulates, a process called deposition.

The final site of accumulation can be any number of places, including on an ancient or modern land surface, such as in sand dunes and on river floodplains, or at the bottom of ancient or modern water bodies, such as an ocean, lake or lagoon. Geologists call these final sites "depositional environments." So, for example, the depositional site of a floodplain is called a "floodplain depositional environment." The Denver Museum of Nature and Science has an excellent book titled "Ancient Denvers"[284] that has artists' renditions of the many landscapes or depositional environments in the Denver area through geologic time. Landscape pictures help you visualize what things looked like in the long-ago past.

As sediment builds up layer on layer, the bottom layers get buried deeper and deeper. The particles get pushed closer together (compacted), and circulating groundwater containing dissolved minerals cements the grains together. As this happens, the sediment becomes sedimentary rock.

Golden's sedimentary rocks consist of clastic and biologic rocks, and, rarely, chemical rocks. Clastic rocks are classified by the size of the particle (sediment), ranging from extremely tiny clays particles to huge boulders. Golden has them all. Using the diameter of a human hair, about 100 microns[285] in diameter, and a pinhead, about 1500 microns in diameter, for comparison:

- clay particles are less than 8 microns;
- silt particles are between 8 and 62 microns;
    - clay and silt, combined, are called "mud:" they are just too tiny to see with the naked eye, so are lumped together.
- sand particles are 62 to 2000 microns
- conglomerate particles range from 2000 microns to more than ten feet in diameter
    - conglomerate particles are further divided into gravel (smallest), pebbles, cobbles, and boulders (largest).

Biologic rocks are classified by their mineral composition and fossil types, such as calcium-carbonate ($CaCO_3$) that forms limestone or organic carbon (C) from fossil plants that forms coal. Golden has both limestone and coal as part of its geology. Chemical sedimentary rocks, such as halite (salt, or sodium chloride) and gypsum (calcium sulfate), form directly out of water by chemical reactions. Golden has smalls amounts of gypsum in one of its sedimentary rock formations.

Sedimentary and Structural Geology: There are Rules!

A couple of important observations about sedimentary rocks have become geological rules, or laws. The first is pretty obvious: as sedimentary rocks build up layer-upon-layer, the oldest rocks are at the bottom of the stack. This principle is called the "Law of Superposition." The second observation is that sediment initially gets laid down across the Earth's surface in pretty much flat layers, a principle is known as the "Law of Original Horizontality." Nicholas Steno, a Dane, figured out these laws around

---

[283] Fossil comes from the French fossile, in turn from the Latin fossus, (to be dug up) the past participle of fodere (to dig). Love those past participles.
[284] Johnson and Raynolds, 2006.
[285] Micron is an abbreviation for micrometer. One micron = 0.000039 inches and 2000 microns is 0.078 inches.

1650. Both have large implications for the field of structural geology, which, in turn, is especially important to the geology of Golden.

Another fundamental geologic principle is known as the "Law of Fossil Succession" and was figured out in the early 1800s. This principle is based on the observation that fossils can be used to determine the age of a rock.[286] For example, different dinosaurs roamed the Earth at very different times. Stegosaurus dinosaur bones (the Colorado State Fossil), are only found in Jurassic rocks, which is much older than Tyrannosaurus rex (T. rex) bones found only in younger Cretaceous rocks. Thus, if you find Stegosaurus bones, then you know the age of the rocks, which in this case is Jurassic (see Geologic (Deep Time) below). Enough to say here, that the geology of Golden was internationally important to figuring out some of these fossil relationships.

Geologic Structures

As we said, sedimentary rock layers, or strata, start out flat. However, originally flat layers can later be tilted by large geologic forces, as is the case for the rocks on the west side of Golden. Tilted rocks indicate much larger features called geologic structures. Essentially there are two types of geologic structures: folds and faults, both of which indicate a mountain-building period (see Rock Cycle below).

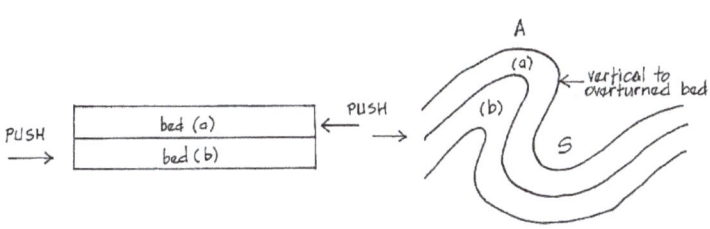

Folds are bends in the rocks in which the layers are bent up as an anticline or down as a syncline (Fig. P1). These two fold types occur together: what goes up, must also go down. Layers in folds can be tilted to various angles,

Fig. P1. Schematic of how flat rock layers can be folded into variably tilted beds and an anticline (A) and syncline (S).

including vertical. Sometimes layers can be overturned where the beds are turned upside down. You can do an experiment at home with a small towel: push it together at the short sides and see how the towel forms folds (a series of anticlines and synclines) as a result. The more you push the short edges together, the bigger the folds. They may flop over, like being overturned. And, yes, rocks actually can be as weak as a towel when they are deep underground. Folds like this are important in Golden's geology.

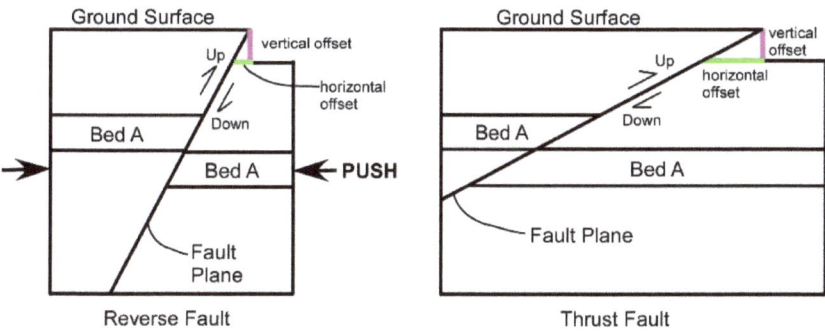

Faults are breaks in rocks across zones as thin as 1/8-inch wide, known as fault planes, where breakage and slippage are concentrated. There are three basic fault types: reverse, normal and strike-slip.[287] In Golden, the faults are nearly all some type of reverse fault (Fig. P2), in which rocks have been pushed up and over their neighbors, along a tilted (or dipping[288]) fault plane. Where the fault-plane tilt is a high angle,

Fig. P2. Reverse (left) and Thrust (right) faults. The rocks on the left side of both faults have been pushed UP. Vertical offset (pink line) is the same in both examples, but horizontal offset (green line) is larger with the thrust fault.

---

[286] For a good discussion see https://pubs.usgs.gov/gip/fossils/succession.html
[287] https://www.usgs.gov/faqs/what-a-fault-and-what-are-different-types?qt-news_science_products=0#qt-news_science_products As usual the USGS has a good explanation of faults.
[288] Geologists use the term "dip" to indicate the tilt of beds. Dip has two properties: angle of "tilt" and orientation, such as how many degrees north, south, east, or west. In this book we try to use "tilt" where possible.

the fault is simply called a "reverse" fault. Where the tilt is less than 45°, the fault is called a "thrust" fault. Both faults are caused by lateral pushing, or compression, just like folds.

Fault offset describes how far a given layer moves up, down, or horizontally along a fault plane. Reverse faults have both horizontal and vertical offset (Fig. P2). Thrust faults have larger horizontal offsets than reverse faults due to the tilt of the fault plane (simple trigonometry). Offset can range from several inches to hundreds of miles. The larger the offset, the more that the fault movement affects the surrounding rocks. Thus, with large fault offsets you might expect that it would actually be a zone or system, of many related fault planes, as is the case with the Golden Fault System.

Folds and faults work together, especially in highly squashed places like the west side of Golden and in most mountain ranges around the world. Forces that change the tilt of the originally flat sedimentary layers into folds and faults are caused by huge Earth, or tectonic, forces, which break, stretch, and/or compress the entire crust over huge areas. The presence of these structures indicates a mountain-building episode. Tectonic forces have operated over the entire Western United States at various times in geologic history, creating mountain-range uplifts, or mountain-building, and the adjoining down-dropped basins that fill with sediment eroded from the uplifted mountains. And, large-scale folding and faulting is part of the geology of the west side of Golden and the Rocky Mountain Front Range: more on that in the book. But, what causes such forces?

The Rock Cycle and Plate Tectonics

All sedimentary and metamorphic rocks, as we discussed above, form from pre-existing rocks. They get changed by heat and pressure. They get eroded, deposited, and buried. Sometimes they get completely re-melted to become igneous rocks. In fact, rocks get recycled over and over: a process called the "rock cycle." Geologic structures (folds and faults) play a key role in the recycling process, as they uplift rocks into mountain ranges as in mountain-building episodes. Those uplifts are then eroded during weathering, with the resulting sedimentary particles ending up being deposited as sediment into down-folded sedimentary basins. Eventually, sediment gets buried deeply enough that it becomes rock. At greater burial depths, it eventually may change by heat and pressure into metamorphic rocks. Underground-melting makes intrusive igneous rocks that, in turn, rise from great depths to the surface where they can be extruded as extrusive volcanic rocks. Volcanoes, in turn, erode to become sedimentary rock particles, and so on and so on in the rock cycle. But why and how does the recycling occur?

Since the 1960's, a huge revolution in geologic thought, called "plate tectonics,"[289] has come about. Geologists now understand that the Earth's crust is formed of rigid plates that slowly slide around on the underlying mantle. At places, the plates are colliding, other places splitting apart, and at some places just passing by. The plates are all interconnected, so what happens at one spot affects everywhere else. And, we now understand that mountain-building episodes are due to interactions at plate boundaries.

Where plates collide, huge compressive, or pushing, forces are transmitted laterally through the crust. In some places, a plate dives down below an overlying plate, a process called subduction. As the rocks on the diving plate go down into the mantle, they get metamorphosed and in some cases are melted to become magma, which when cooled becomes intrusive igneous rock. Where magma forms and rises to the Earth's surface above a subducting plate, volcanoes are common, like along the West Coast of the U.S. in Oregon and Washington and the Aleutian Islands in Alaska.

Where plates split or rift apart, huge extensional, or pulling, forces are transmitted laterally across the crust to form rift zones. As the crust thins across rift zones, hot magma from the lower crust and upper mantle rises to the surface in the form of volcanoes, like in the Rio Grande Valley of New Mexico.

---

[289] For much more detail see https://en.wikipedia.org/wiki/Plate_tectonics

Places where plates pass by horizontally create "transform" faults, a type of "strike-slip" fault, like the San Andreas Fault in California where Los Angeles (on the west side of the fault) and San Francisco (on the east side of the fault) are growing farther and farther away from eachother at a rate of up to 2 inches per year. For geologists, that is a pretty fast rate. When the San Andreas Fault moved abruptly in the 1906 San Francisco earthquake, fences near the maximum rupture were offset 21 feet laterally.[290] Fortunately, Colorado is located far from this active geologic feature.

Earthquakes, volcanoes, the creation of magmas, mountain-building episodes (orogenies), and the creation and destruction of rock types (the rock cycle) are all explained by Earth's tectonic plates moving around.[291] No wonder that plate tectonics became the unifying principle for understanding all aspects of geology. Although too large a topic to delve into in this book, the website hyperlink in the footnote at the bottom of the previous page will lead you to more resources about Plate Tectonics.

Geologic (Deep) Time

Before the 1930s, geologists spoke about geologic time without knowing the actual amount of time represented by the rocks. Geologists figured out the relative ages of rocks by going around the world and piecing together the order of sedimentary rock layers (i.e. Superposition: the oldest layer on the bottom) by using the typical fossils found in those rocks (the Law of Fossil Succession). As the layer and fossil orders became understood, they turned into a geologic time scale with names given to major groups of time, as represented by changes in the fossil record.

The geologic time scale is divided into blocks of time called Eras and Periods,[292] all based on groups of fossils that are present only during those times, like the bones of Stegosaurus and T. rex. The boundaries between Eras (Paleozoic, Mesozoic, and Cenozoic) are times when mass extinctions of plants and/or animals occur in the rock record. For example, 90% of all marine life went extinct at the end of the Paleozoic Era, and famously, dinosaurs and many forms of marine life went extinct at the end of the Mesozoic Era. Remarkably, both of these geologic boundaries are present in Golden, creating part of a fascinating story of important research and discovery. Eras usually were named for the perceived antiquity of life forms. Thus "Mesozoic" means "middle animal life." Can you guess what Paleozoic means?[293]

Periods are smaller time intervals within Eras and record less-massive extinctions. Periods were named after the place on Earth where the layers and/or fossils representing them were first defined. So, for example, the "Cretaceous Period" was named for the fossil organisms that made chalk, in the White Cliffs of Dover in England. In Greek, "creta" means "chalk." As another example, rocks of the Jurassic Period were first found in the Jura Mountains of Switzerland.

But what happened after the 1930s? What was so special about that time? That is when scientists discovered isotopes, which are versions of the same element with the same number of protons but with differing amounts of neutrons. The difference in neutrons causes subtle differences in the mass of that element, which can be measured. Radioactive elements, in particular, throw off neutrons in a controlled manner known as radioactive decay. Measuring the amounts of particular isotopes in the elements composing a mineral yields the age of the rock.[294]

Since the 1930s, radiometric age-dating of minerals in volcanic ash layers, lava flows, granite, and gneiss, made it possible to assign actual ages to rocks and fossils and to the time boundaries between geological Periods and Eras. Thus from radiometric dating, geologists know that the unconformity between the Fountain Formation and the underlying basement rocks (gneiss) represents about 1360 million years (Chapter 2).

---

[290] A good explanation of the San Andreas Fault at https://pubs.usgs.gov/gip/earthq3/safaultgip.html
[291] A good graphic of the rock cycle at https://www.geolsoc.org.uk/ks3/gsl/education/resources/rockcycle.html
[292] The complete official time scale is at https://www.geosociety.org/documents/gsa/timescale/timescl.pdf
[293] Paleo means "ancient" and "zoic" means "animal life" from Latin. Similarly, Cenozoic means new animal life.
[294] For a detailed discussion of radiometric age-dating see https://en.wikipedia.org/wiki/Radiometric_dating

The millions of years of time represented by the geologic time scale, as we know it today, is referred to as "deep time." In conversations, geologists casually discuss millions of years like they are months in the year. In reality, mentally grasping millions of years of Earth and planetary history is hard, just like imagining millions of miles in space. Astronomers get around that issue by talking about distances in units of "parsecs," in which one parsec is 19 trillion miles. Thus, discussing a few hundred parsecs does not sound so intimidating, unless you do the math. Similarly, the easiest way to think about millions of years is convert them in your mind to a new unit such as "million-years," as suggested by Walter Alvarez.[295] In this book we will simply call that time unit: "Ma."[296] Saying 65 or 300 Ma does not sound so overwhelming.

Unconformities in Geology

An important follow-up to the Law of Fossil Succession deals with missing fossil zones in a stack of sedimentary rocks. Missing fossil zones are evidence of erosion or non-deposition, which in turn represents a gap in the rock record: i.e. deposition was interrupted for some reason. That gap, commonly referred to as a "time gap,"[297] is called an unconformity. The gap is like having pages missing in your favorite book: a very dissatisfying experience! You can imagine that a with only a few pages missing, you could figure out the thread of the story. But with entire chapters missing, you might be completely lost. In the Golden area, most of the rocks and fossils representing the Paleozoic Period are missing at the contact between the sedimentary rocks of the Fountain Formation and the underlying very old metamorphic gneiss of the Front Range, forming what is known as the "Great Unconformity." A lot of "pages" are missing at that contact. By contrast, the contact between the Laramie Formation and the overlying Arapahoe Formation is an unconformity of shorter duration, perhaps less than 500,000 years, representing only a few "missing pages" in the book.

A Philosophy of the Rock Record

Collectively, the sum of all geology is the "rock record." The concept is similar to reading all of the books in a library to get the sum of all geologic knowledge. But in geology those "books" are rocks with all their groupings and variations. A deep-rooted knowledge of geology is an ability to read the rocks. An interesting point about the rock record is its four-dimensionality, an aspect also shared by astronomy and astrophysics. That is, geology deals with the map view of the landscape (the first two dimensions of x and y), what is under the landscape or below the ground (the third dimension of z), and how that changes through time (the fourth dimension of t). Yes, geologists "see" below the ground across a landscape, and they see it through the vast amount of geologic time.

The Present and the Past: Uniformitarianism and Catastrophism

Geologists are taught that the "present is the key to the past." The statement, popularized by Charles Lyell in 1830 but originated by James Hutton (1726-1797),[298] explains the features of the Earth's crust by means of familiar natural processes, like water and wind erosion and deposition, over geologic time. The principle is known as uniformitarianism, and has two major implications.

The first implication is the fact that landforms and rocks visible in the landscape today, like rivers, beaches and their deposits, also formed in the geologic past in the same way as they do today. That is, the Earth processes that one sees today follow the laws of physics and chemistry and have always operated the same way over time. For example, 18th and 19th century observers found that huge boulders occurring near glaciers in the Alps were actually dropped out of melting glacial ice. They also

---

[295] From Alvarez, W., 1997, "T. Rex and the Crater of Doom," p. 28
[296] Ma is the abbreviation for mega-annum, which means million years in Latin.
[297] Gaps in time are physically impossible, but the phrase, "time gap," is commonly used.
[298] A complete biographical sketch of Hutton and his observations: https://en.wikipedia.org/wiki/James_Hutton

observed that valleys carved by then-active glaciers in the Alps[299] have a distinctive U-shaped cross section. Going back to Great Britain and observing out-of-place boulders[300] lying in U-shaped valleys, these geologists concluded that the features were formed by long-ago melted glaciers. They also realized that Great Britain and the rest of the Earth had been subjected to great Ice Ages in the past. While that conclusion may seem obvious to us today, it was not obvious 200 years ago.

Such logic is used over and over in understanding how landforms and rocks are formed. For example, sedimentary geologists go to the Bahamas to see how limestone rocks are forming today, as a way to understand ancient limestone. Other geologists travel to Oman and the island of Crete to study the metamorphic and intrusive igneous rocks formed in the mantle that today are uplifted and visible on the surface. No wonder geologists, as a group, like to travel around the world. As one famous geologist said, "Never turn down a free plane ticket."[301]

The second implication of uniformitarianism is that Earth processes (erosion, deposition, and mountain-building) are gradual, or uniform. They occur over a long period of time that is much longer than hundreds of generations of human existence. In the 18th century a large controversy existed whether or not Earth processes were gradual, as opposed to the competing philosophy of "catastrophism" in which processes were short, within days or weeks. Since the 1950s though, as more and more data have been collected from the deep oceans, high mountains, and our solar system (e.g., a comet hit Jupiter in 1994), our understanding of Earth and other planetary processes has grown exponentially. Geologists have come to realize that some things in long-ago (deep) time have no modern counterparts and that catastrophes are indeed part of the rock record, as we will see in the geology of Golden. Perhaps a better statement is that the "present is a key to the past" and a powerful key at that.[302]

Reading Golden's Geologic Map

Geologic maps were first made in early 1800s. The story of the first geologic map of an entire country is told by Simon Winchester in his book, "The Map that Changed the World." It is an amazing tale of an English engineer, William Smith (1769-1839), who figured out how to put rock types on a map to predict where coal seams might be found underground in England. Smith's trials and tribulations, including getting thrown into debtors prison and having to sell his rock collection to the British Museum (it is still there in a glass case), is a compelling story of human nature. Most geologists have read that book at least once. It is a good read whether you are a geologist or not.

Geologic maps are a lot different from simple road maps. The geologic map of the Golden area (Fig. P3) shows a spectrum of colors and symbols that represent the ages and distributions of geologic formations. In the background of the map are streams, rivers, and lakes in light blue and light-gray

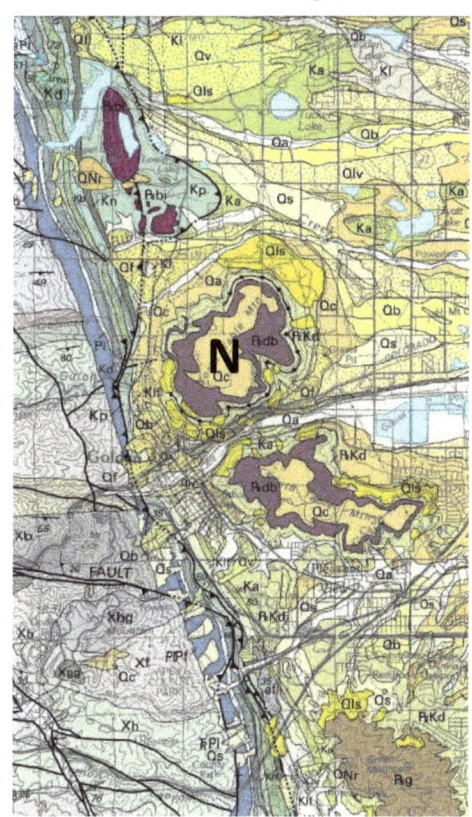

Fig. P3. Geologic map of Golden area by Kellogg et al., 2008. Colors, lines, and symbols discussed in text. "N" is North Table Mountain. Kellogg et al., 2008.

---

[299] This was during the Little Ice Age, which ended around 1900. In the early 1800s advancing glaciers in the Alps threatened entire villages, creating great consternation. The Little Ice Age ended as climate warmed.
[300] The out-of-place boulders are called "glacial erratics."
[301] A quote from Warren Hamilton (1925-2018), a true global geologist.
[302] An excellent discussion of this and other issues is in Marsha Bjornerud's book, "Timefulness: How thinking like a geologist can help save the world," published in 2018.

lines representing highways, streets, topographic contours, and public land survey boundaries. All in all, it is a visually "busy" map that portrays a lot of information.

Colors and letter symbols on a geologic map are standardized so geologists immediately know the ages of the rocks. For example, different shades of green represent Cretaceous rocks, and the letter-symbol for the Cretaceous Period always begins with the letter "K," such as Klf for Laramie and Fox Hills, grouped together. Igneous rocks also have distinct map colors. Paleogene-age extrusive (above ground) rocks, like lava flows, are dull brown and labeled PEdb,[303] whereas intrusive igneous rocks, like those of old volcanoes, are red-brown and labelled PEbi. Yellow and orange represent the youngest Quaternary surface deposits that lie on top of the older bedrock. Symbols for Quaternary units always begin with the letter "Q." For example, on North Table Mountain (labelled "N" on Fig. P3), you can see the dull brown lava labelled as PEdb has a covering of more recent colluvium (tan-colored Qc) on top of the lava in the center of the mountain. Also, on the steeper sides of the mountain, below the dull brown lava, you see Denver Formation (light green PEKd) and lower down the slope are some recent landslide deposits (yellowish Qls).

Heavy black lines of various types (solid, dotted) on the map (Fig. P3) show the locations of faults and lava flows in the Golden area. Prominent curved black lines, some with small triangles, represent fault lines of various types. At North Table Mountain ("N" on Fig. P3), a black line with heavy black dots represents lava flows cropping out around the sides of that mountain.

Accordingly, at any location on the map, you can tell the rock formation and its age at that spot by the color and letter symbol. The rock types, like sandstone or gneiss, can be figured out by using the legend on the stratigraphic column (geologic "Rosetta stone;" Fig. P4). If you are lucky (?) you might find that your location is on a fault line separating two rock formations. Of course, there is also "an app for that" in the form of **Rockd** on your phone or tablet.[304]

Reading Golden's Geologic Column

The geologic column for Golden (Fig. P4), also known as a stratigraphic column, is a geologic "Rosetta Stone" that summarizes the information on a geologic map such as rock ages and formations, and provides additional information about rock types and formation thicknesses. A geologic

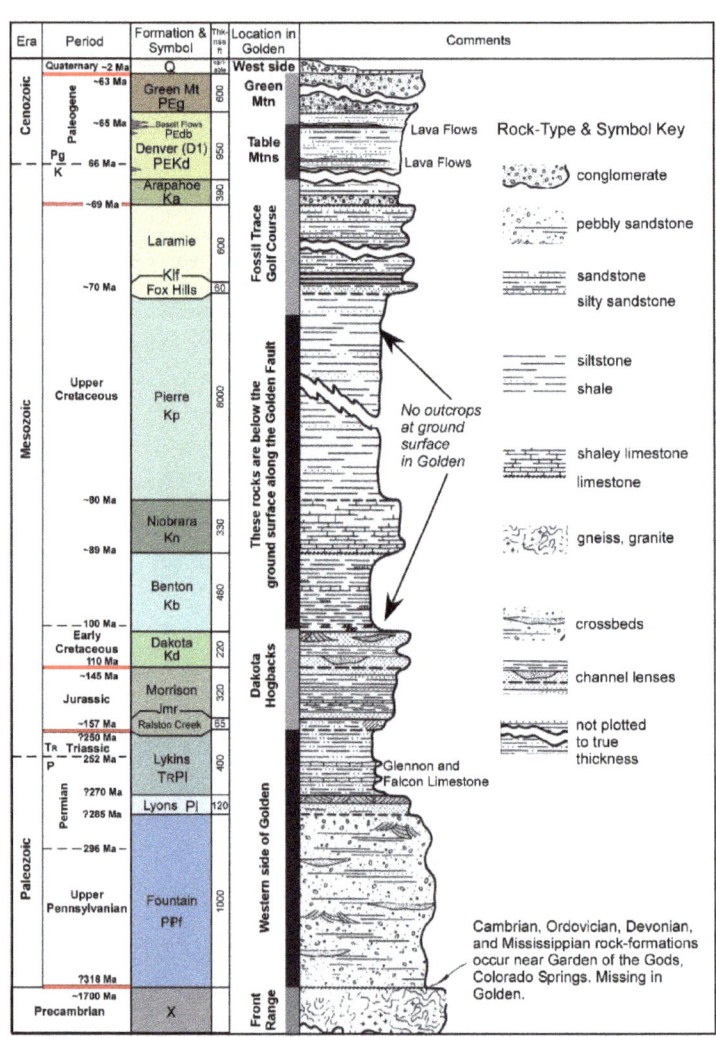

Fig. P4. Geologic column, or geologic "Rosetta Stone" for Golden, discussed in text.

---

[303] PEdb means: Paleogene (PE) Denver Formation basalt; PEbi means Paleogene basalt-intrusive; PEKd means Paleogene-Cretaceous Denver Formation.

[304] **Rockd**, created by the Macrostrat Lab at the University of Wisconsin–Madison: https://macrostrat.org/ is a great app, and the geologic map for Golden is pretty accurate.

column shows a stack of rock layers, or strata, arranged with the oldest at the bottom and the youngest at the top. The column height is scaled to the thickness of the rock layers. Note that several units are not plotted to true thickness. For example, the 8000-foot thick Pierre Shale has a zigzag break in the middle. If it were plotted correctly, the Pierre Shale, alone, would take up an entire page by itself, making the whole column into several pages. For convenience, it and four other formations are not plotted to true thickness.

The far-left side of the column shows the geologic time-names and ages (in Ma) associated with the rock layers. Note that in places a time boundary has a single age (like the K-Pg boundary at 66 Ma) and in other places ages are listed above and below a boundary), such as at the Quaternary-Paleogene boundary. The two ages represent the age at the top of the Green Mountain Formation at 64 Ma and at base of the next higher formation, the Quaternary at about 2 Ma. Therefore, because about 62 Ma of time is not represented by rock, the Quaternary-Paleogene boundary is an unconformity. Five unconformities with large to small time gaps are present in Golden. They are shown as horizontal red lines in the "Period" column on Fig. P4.

The "Formation and Symbol" column shows names for the rock layers, grouped into formations[305] with different colors and letter symbols for each named formation. Formation names indicate the place where geologists first described the best example of a particular rock unit. For example, the Fountain Formation is best exposed at Fountain, Colorado. Where do you think that the name of the Morrison Formation comes from? How about the Lyons Formation and Denver Formation?[306] Make a guess. Sometimes it is pretty easy.

In Golden, as elsewhere in northeastern Colorado, the rock/formation layers stack up to thousands of feet of rock. The thickness of each rock unit, excluding the Precambrian (Unit X), is listed next to each formation. Adding up all the thicknesses of the formations between the Precambrian and Quaternary equals 13,525 feet. For example, say you drilled a vertical hole starting at top of Green Mountain. If you drilled to 13,525 feet below the ground surface you would reach Precambrian rocks, going through all the formations on the geologic column on the way down.

In the center is a column ("Features in Golden") with colored vertical bars. These are our groupings of formations that correspond to the major landforms in Golden, all of which we will discuss in this book. For example, the feature with a tan-colored bar, labelled "Table Mtns" corresponds to prominent landforms of North and South Table Mountains in Golden, which are composed of the Denver Formation. Similarly, the feature with the gray vertical bar represents the prominent western side of Golden, the Front Range of the Rocky Mountains, which is composed of Unit X, Precambrian gneiss.

On the far-right side of the column is symbol key to various rock types that match the formation names. Rock types listed in the descriptions are the same as those discussed in the "Three Rock Types" section at the beginning of this Geology Primer.

In closing, this primer is meant to give you a flavor of some of the main concepts used in this book. There are many web resources about geology, and if you truly are motivated, take a beginning geology course at your nearest community college!

---

[305] A geologic "formation" is formally defined as group of rocks with distinct composition that are mappable over a large area.
[306] Dakota Formation, Dakota County, NB; Laramie Formation, Laramie, WY; Pierre Formation, Pierre, SD; and so forth.

References Cited

Abbott, D., D.A. McCoy, and R.W. McLeod, 2007, Colorado Central Railroad: Sundance Publications, Ltd., Denver, Colorado, 416 p.

Alderson, V.C., 1912, The experimental ore dressing and metallurgical plant of the Colorado School of Mines: Metallurgical and Chemical Engineering, v. 10, no. 5, p. 269-272.

Alvarez, W., 1997, T. Rex and the crater of doom: Princeton University Press, New Jersey, 208 p.

Amuedo, C.L., A.R. Myers, R.A. Lindvall, J.L. Hynes, and J.B. Ivey, 1978, Coal and clay mine hazard study and estimated unmined coal resources, Jefferson County, Colorado, prepared for the County of Jefferson and the cities of Arvada, Golden and Lakewood, CO: unpublished report by Amuedo and Ivey (consulting firm) for the Jefferson County Planning Department, approximately 92 p. Arthur Lakes Library, Mining History collection, CSM.

Arbrogast, B., D.H. Knepper, Jr., R.A. Melick, and J. Hickman, 2002, Evolution of the Clear Creek corridor, Colorado—Urbanization, aggregate mining, and reclamation: USGS Geologic Investigations Series I-2760, 41 p.

Argall, G.O., Jr., 1949, Clay in Colorado: Colorado School of Mines Quarterly, v. 44, no. 2, p. 89-109.

Arthur Lakes Library, CSM, Colorado Digitization Project of the Russell L. and Lyn Wood Mining History Archive https://mountainscholar.org/handle/11124/9853 "tunnel house" Last accessed 18 June 2021.

Baker, V.R., 1974, Paleohydraulic interpretation of Quaternary alluvium near Golden, CO: Quaternary Research, v. 4, p. 94-112.

Baskin, O.L., and N. Millet, 1880, History of Clear Creek and Boulder Valleys: Baskin and Company, Chicago, 713 p.

Benson, R.D., 2016, Protecting river flows for fun and profit: Colorado's unique water rights for Whitewater Parks: Ecology Law Quarterly, v. 42, p 753-786.

Berthoud, E.L., 1872, Tusk of an elephant or mastodon found in Colorado: American Journal of Science, Series 3, v. 3, p. 302.

Berthoud, E.L., 1880, History of Jefferson County, in Baskin, O.L, and N. Millet, History of Clear Creek and Boulder Valleys: Baskin and Company, Chicago, p. 353-378.

Bjornerud, M., 2018, Timefulness: How thinking like a geologist can help save the world: Princeton University Press, New Jersey, 224 p.

Borowsky, L., 2015, Home is where the mine is: Mines Magazine, p.22-23. https://minesmagazine.com/10020/ Last accessed 15 June 2021.

Brannon Sand and Gravel Company, 1979, Application for rezoning: mining and reclamation permit: Pit 24, Phase 1, Clear Creek and Phases 2, 3, and 4, Golden Gate Canyon: Jefferson County, unpublished report, 528 p.

Brown, R. W., 1943, Cretaceous-Tertiary boundary in the Denver Basin, Colorado: Geological Society of America Bulletin, v. 54, p. 65-86.

Butler, W.B., 1992, Archeological survey of Camp George West and the Works Progress Administration South Table Mountain basalt quarries, Jefferson County, Colorado; Report produced for the Colorado National Guard, Colorado State Archaeological Permit Number 92-5: National Park Service, Interagency Archeological Services, 59 p.

Carder, C., 2007, From landfill to athletic fields: Building waterless athletic fields atop of Rooney Landfill in Jefferson County, Colo.: Building, Design and Construction website: https://www.bdcnetwork.com/landfill-athletic-fields Last accessed 8 June 2021

Carpenter, K., and D. B. Young, 2002, Late Cretaceous dinosaurs from the Denver Basin: Rocky Mountain Geology, v. 37, p. 237-254.

Carpenter, K., and E. Lindsey, 2019, Redefining the Upper Jurassic Morrison Formation in the Garden Park Natural Landmark and vicinity, eastern Colorado: Geology of the Intermountain West, Utah Geological Association, v. 6, p. 1-30.

Carroll, C.J., and M.A. Bauer, 2002, Historic coal mines of Colorado: Colorado Geological Survey Information Series 64. https://coloradogeologicalsurvey.org/publications/historic-coal-mines-colorado/ Last accessed 15 June 2021

Cather, S.M., C.E. Chapin, and S.A. Kelley, 2012, Diachronous episodes of Cenozoic erosion in southwestern North America and their relationship to surface uplift, paleoclimate, paleodrainage, and paleoaltimetry: Geosphere, v. 8, p. 1177-1206.

Chapin, C.E., S.A. Kelley, and S.M. Cather, 2014, The Rocky Mountain Front, southwestern USA: Geosphere, v. 10, p. 1043-1060.

Cole, J.C., and W. A. Braddock, 2009, Geologic map of the Estes Park 30' x 60' quadrangle, north-central Colorado: USGS Scientific Investigation Map 3039, with pamphlet. Map scale 1:100,000.

Colorado Division of Homeland Security and Emergency Management, 2021, http://www.coemergency.com/2010/01/colorado-earthquake-information.html Last accessed 13 October 2021.

Colorado Encyclopedia, 2021, Ku Klux Klan in Colorado: https://coloradoencyclopedia.org/image/ku-klux-klan-denver Last accessed 14 October 2021.

Colorado Mining Gazette, 1885, as cited in footnotes. Available at https://www.coloradohistoricnewspapers.org/

Colorado Transcript (became the Golden Transcript in 1969), as cited in footnotes. Available at https://www.coloradohistoricnewspapers.org/

Colorado Water Institute, 1952, A hundred years of irrigation in Colorado, 1852-1952: Colorado Water Conservation Board and Colorado Agricultural and Mechanical College, Fort Collins, CO, 111 p.

Coors Porcelain, 1935, Lump of Clay: Coors Porcelain, Golden Colorado, 56 p. (corporate magazine)

Cross, W.A., and W.F. Hillebrand, 1882, Communications from the U.S. Geological Survey, Rocky Mountain Division. I. On the minerals, mainly zeolites occurring in the basal of Table Mountain, near Golden, Colorado: American Journal of Science, Third Series, v. 23, p. 452-458.

Cross, W.A., and W.F. Hillebrand, 1885, Contributions to the mineralogy of the Rocky Mountains: USGS Bulletin 20, 115 p.

Cross, W.A., 1896, Denver Formation, in S.F. Emmons et al., Geology of the Denver Basin in Colorado: U.S. Government Printing Office, p. 155-254.

Dames and Moore, 1981, Final report, geologic and seismologic investigations for Rocky Flats Plant: U.S. Department of Energy: v. 1, no. 10805-041-14. (Available at Arthur Lakes Library, CSM).

Dechesne, M., R.G. Raynolds, P.E. Barkmann, and K.R. Johnson, 2011, Notes on the Denver Basin geologic maps: Bedrock geology, structure and isopach maps of the Upper Cretaceous to Paleogene strata between Greeley and Colorado Springs, Colorado: Colorado Geological Survey, Department of Natural Resources, 35 p. and GIS data.

Denver Post, 2021, See where 1920s Denver Ku Lux Klan members lived, search the database: https://www.denverpost.com/2021/09/14/denver-kkk-members-map-database/ Last accessed 14 October 2021.

Drewes, H., 2008, Table Mountain shoshonite porphyry lava flows and their vents, Golden, Colorado: USGS Scientific Investigations Report 2006-5242, 28 p.

Drewes, H., and J. Townrow, 2005, Trailwalker's guide to the Dinosaur Ridge, Red Rocks and Green Mountain area: 2nd edition, Friends of Dinosaur Ridge, Morrison, Colorado, 86 p. Map scale 1:18,000.

Eckley, W., 2004, Rocky Mountains to the world: A history of the Colorado School of Mines: Donning Company Publishers, Virginia Beach, Virginia, 224 p.

Emmons, S.F., W. Cross, G.R. Eldridge, 1896, Geology of the Denver Basin in Colorado: U.S. Government Printing Office: 556 p. and Plates (maps).

Fielding, C.R., T.D. Frank, S. McLoughlin, V. Vajda, C. Mays, A.P. Tevyaw, A. Winguth, C. Winguth, R.S. Nicoll, M. Bocking, and J.L. Crowley, 2019, Age and pattern of the southern high-latitude continental end-Permian extinction constrained by multiproxy analysis: Nature Communications, v. 10, no. 385, 12p.

Follansbee, R., and L.R. Sawyer, 1948, Floods in Colorado: USGS Water Supply Paper 997, 151 p.

Foothills Genealogical Association of Colorado, 1993, Jefferson County Colorado, History of Golden, WPA Project 3584, over 800 p. https://www.foothillsgenealogy.org/page-18271 Last accessed 15 June 2021.

Gardner, R., 2004, Steamboats on Clear Creek: Historically Jeffco, Issue 25, p. 26-29.

Gardner, R., 2015, Magic Mountain–Destination: Fun and fantasy! Historically Jeffco, Issue 36, p. 2-10.

Gardner, R., 2017, Sesquicentennial sites mark momentous time for Jeffco: Historically Jeffco, Issue 38, p. 25-27.

Gardner, R., 2020, Green gold sowed in Golden: Historically Jeffco, Issue 41, p. 24-25.

Geijsbeek, S., 1901, The Colorado clay deposits: The Clay Worker, v. 36, p. 424-426.

Golden Globe newspaper, as cited in footnotes. Available at https://www.coloradohistoricnewspapers.org/

Golden History Museum Epic Events, 2021, Timeline: https://www.goldenhistory.org/wp-content/uploads/2016/08/EpicEventsTimelineGuide_web.pdf Last accessed 15 June 2021.

Golden Landmarks Association, various dates: http://goldenlandmarks.com/brickyard-house/ ; http://goldenlandmarks.com/brickyard-history/ ; http://goldenlandmarks.com/history-of-the-pullman-house/ ; http://goldenlandmarks.com/history-of-the-pullman-house/ Last accessed 17 June 2021.

Grunska, J., 2003, Traceries: Historically Jeffco, Issue 24, p. 2-6.

Hagadorn, J.W., K.R. Whitely, B.L. Lahey, C.M. Henderson, and C.S. Holmes-Denoma, 2016, The Permian-Triassic transition in Colorado, in Unfolding the Geology of the west: Geological Society of America Field Guide 44, p. 73-92.

Hayden, F.V., 1869, Preliminary Field Report of the U.S. Geological Survey of Colorado and New Mexico: Government Printing Office, Washington, D.C., 155 p.

Higley, D., and D.O. Cox, 2007, Oil and gas exploration and development along the Front Range in the Denver Basin of Colorado, Nebraska, and Wyoming, in Higley, D., Petroleum Systems and assessment of undiscovered oil and gas in the Denver basin province, Colorado, Kansas, Nebraska, South Dakota, and Wyoming: USGS Province 39: USGS Digital Data Series DDS-69-P, Chapter 2, 41 p.

History Colorado, Ku Klux Klan Ledgers, Greater Denver area, 1920s, Ledger 2 page 46: https://www.historycolorado.org/kkkledgers Last accessed 14 October 2021.

History Colorado, National and State Register, Listed Properties: 2021: https://www.historycolorado.org/national-state-register-listed-properties Last accessed 13 October 2021,

Houck, K.J., M.G. Lockley, M. Caldwell, and B. Clark, 2010, A well-preserved crocodilian trackway from the South Platte Formation (Lower Cretaceous), Golden, Colorado, in Crocodyle tracks and traces, New Mexico Museum of Natural History and Science, Bulletin 51, p. 115-120.

Hubert, J.F., 1960, Petrology of the Fountain and Lyons formations, Front Range, Colorado: Colorado School of Mines Quarterly, v. 55, p. 1-242.

Iddings, J.P., 1895, Absarokite-Shoshonite-Banakite series: Journal of Geology, v. 3, no. 8, p. 935-959.

Jackson, W.H., 1872, Golden City, Jefferson County, Colorado, 1872, in Descriptive Catalog of the Photographs of the United States Geological Survey of the Territories, W. H. Jackson, Photographer, Second Edition, Illustrated, 1872 Series, page 39, Nos. 372, 373. https://library.usgs.gov/photo/#/item/51dc982be4b097e4d383a66d Last accessed 14 June 2021.

Jefferson County Colorado: Jefferson County. Non-county agency manuscript maps of Jefferson County, ca. 1879 (map from ca. 1879 showing south Golden roads and GC&SP graded rail routes) and ca. 1890. Series #116. Available from https://www.jeffco.us/DocumentCenter/View/1090/Manuscript-Maps-of-Jefferson-County-Finding-Aid-PDF Last accessed 18 June 2021.

Johnson, J.H, 1930, The paleontology of the Denver Quadrangle: Proceedings of the Colorado Scientific Society, v. 12, p. 355-378.

Johnson, J.H., and W.A. Waldschmidt, 1930, Famous Colorado mineral localities: Table Mountain and its Zeolites: American Mineralogist, V. 10, p. 118-120.

Johnson, K.R., and R.G. Raynolds, 2006, Ancient Denvers: Scenes from the past 300 million years of the Colorado Front Range: Denver Museum of Nature and Science, Fulcrum Publishing, Golden, Colorado, 34 p.

Kauffmann, E.G., G.R. Upchurch, Jr., and D.J. Nichols, 1990, The Cretaceous–Tertiary boundary interval at South Table Mountain, near Golden, Colorado, in Kauffman E.G., and Walliser, O.H., eds., Extinction Events in Earth History: Lecture Notes in Earth Sciences, v. 30. Springer, Berlin-Heidelberg, p. 365-392.

Kellogg, K.S., R.R. Shroba, B. Bryant, and W.R. Premo, 2008, Geologic map of the Denver West 30' x 60' quadrangle, north-central Colorado, pamphlet to accompany Scientific Investigations Map 3000, 48 p.

Kile, D.E., 2002, Clay minerals of the Front Range: A field guide to the geology, history, and clay mineralogy of the Chieftain mines, Dinosaur Ridge, Patch mine, and other localities along the Front Range from Denver to Boulder, Colorado: USGS Open-file report 02-413, 79 p.

Koch, A.J., D.S. Coleman, and A.M. Sutter, 2019, Provenance of the upper Eocene Castle Rock conglomerate, south Denver Basin, Colorado, U.S.A.: Rocky Mountain Geology, v. 53, p. 29-43.

Kostka, W., Sr., 1973, The Pre-Prohibition History of Adolph Coors Company 1873-1933: Adolph Coors Company, Golden, CO, 106 p.

Langer, C.J., R.A. Martin, C.K. Wood, G.L. Snyder, and G.A. Bollinger, 1991, The Laramie Mountains, Wyoming, earthquake of 18 October 1984: A report on its aftershocks and seismotectonic setting: USGS Open-File Report 91-258, 41 p.

Langer, W.H., and M.L. Tucker, 2003, Specification Aggregate Quarry expansion: A case study demonstrating sustainable management of natural aggregate resources: USGS Open-File Report 03-121, 11 p.

LeRoy, L.W., and R.J. Weimer, 1971, Geology of the Interstate 70 road cut, Jefferson County, Colorado: Professional Contributions of Mines, no. 7, one over-sized sheet.

Lewis, R., 2013, The lime-burning industry: Historically Jeffco, Issue 34, p. 24-28

Lindsey, D.A., W.H. Langer, and D.H. Knepper, 2005, Quaternary alluvium of the Front Range: USGS Professional Paper 1705, 70 p.

Litvak, D., E. Raath, N. VanderKwaak, and J. Wahlers, 2020, Historic streetcar systems of Colorado: CDOT Report202-11, Colorado Department of Transportation Research, Denver CO, 440 p. https://www.codot.gov/programs/research/pdfs/2020-research-reports/historic-streetcar-systems-of-colorado-1/cdot-2020-11.pdf Last accessed 8 June 2021.

Lockley, M.G., and C. Marshall, 2014, Field guide to the Dinosaur Ridge area: Friends of Dinosaur Ridge, Morrison, Colorado, 40 p.

Lockley, M.G., B. Simmons, and G. Daggett, 2014, A new dinosaur track site in the Dakota Group (Cretaceous) of the Golden Area, Colorado, in Fossil footprints of western North America: New Mexico Museum of Natural History Bulletin 62, p. 361-364.

Lockley, M.G., K. Chin, K. Houck, M. Matsukawa, and R. Kukihara, 2009, New interpretations of Ignotornis, the first-reported Mesozoic avian footprints: implications for the paleoecology and behavior of an enigmatic Cretaceous bird: Cretaceous Research, v. 30, p. 1041-1061.

Marvine, A.R., 1874, Report of Arch. R. Marvine, Assistant Geologist directing the Middle Park division, *in* Hayden, F.V., 1874, Seventh annual report of the U.S. geologic and geographical survey of the territories embracing Colorado, a progress of the exploration of the year 1873: U.S. Government Printing Office, Washington, D.C., p. 83-189.

Matthews, N.E., J.A. Vasquez, and A.T. Calvert, 2015, Age of the Lava Creek supereruption and magma chamber assembly at Yellowstone based on $^{40}Ar/^{39}Ar$ and U-Pb dating of sanidine and zircon crystals: Geochemistry, Geophysics, Geosystems, v. 16, p. 2508-2528.

Matthews, V., K. KellerLyn, and B. Fox, eds., 2003, Messages in Stone: Colorado's colorful geology: Colorado Geological Survey SP-52, 163 p.

McKinney, K.C., ed., 2003, Digital archive: Previously unpublished sketches by Henry W. Elliott united with the preliminary field report of the United States Geological survey of Colorado and New Mexico, 1869, by F.V. Hayden: USGS open-file report 03-384, 209 p.

McNeil, J., 1890, Fourth biennial report of the State Inspector of Coal Mines of the State of Colorado, for the years of 1899-90: Collier and Cleaveland Lithographic Company, State Printers, Denver, Colorado, 100 p.

Meldahl, K.H., 2011, Rough-hewn land: A geologic journey from California to the Rocky Mountains: Berkeley and Los Angeles, University of California Press, 296 p.

Mile High Flood District, 2020, History of Floods in the Bear Creek drainage: https://www.udfcd.org/FWP/ebb/bear_history.html Last accessed June 8, 2021.

Milliken, A. E.G., L.E. Morgan, and J. Noblett, 2018, 40Ar/39Ar geochronology and petrogenesis of the Table Mountain Shoshonite, Golden, Colorado, U.S.A.: Rocky Mountain Geology, v. 53, p. 1-28.

Noe, D.C., J.M. Soule, J.L. Hynes, and K.A. Berry, 1999, Bouncing boulders, rising rivers, and sneaky soils: A primer of geologic hazards and engineering geology along Colorado's Front Range, in Colorado and adjacent areas: Geological Society of America Field Guide 1, p. 1-19.

Norman, C., 1996, Golden old and new: A walking tour Guide: Preservation Publishing, Lakewood, Colorado, 84 p.

Obradovich, J.D., 2002 Geochronology of Laramide synorogenic strata in the Denver Basin, Colorado: Rocky Mountain Geology, v. 37, p. 165-171

Oredigger magazine, as cited in footnotes. Available at https://www.coloradohistoricnewspapers.org/

Parfet, W.G. (Bill), Sr., and D. Ryder, 1977, Interview of Bill Parfet by Daryl Ryder, 12 January 1977. Transcription of 30-minute taped interview concerning historic rail lines and clay mines in the Golden area: Object number 2021.046.0001, Colorado Railroad Museum Golden, Colorado.

Parfet, W.G. (Chip), Jr., 2007, History of the Rubey clay mine, Essay by Wm. (Chip) Parfet), *in* R.J. Weimer, Field trip of clay resources and products in the Golden Area, May 24, 2007, unpublished, in the collection of Dr. Robert J. Weimer, 26 p.

Parker, B.H., Jr., 1974, Gold placers of Colorado: Colorado School of Mines Quarterly, v. 69, no. 3, Book 1, 267p.

Patton, H.B., 1904, Faults in the Dakota Formation at Golden, Colorado: Geological Society of America Bulletin, v. 18, p. 26-32.

Powell, J.W., 1876, Biographic notice of Archibald Robertson Marvine: Bulletin of the Philosophical Society of Washington, D.C., p. 53-60. https://archive.org/details/biographicalnoti00powe/page/n1/mode/2up Last accessed 14 June 2021.

Raynolds, R. G., 2002, Upper Cretaceous and Tertiary stratigraphy of the Denver Basin, Colorado: Rocky Mountain Geology, v. 37, p. 111-134.

Raynolds, R.G., and K. Johnson, 2003, Synopsis of the stratigraphy and paleontology of the uppermost Cretaceous and lower Tertiary strata in the Denver Basin, Colorado: Rocky Mountain Geology, v. 38, p. 171-181.

Reichert, S.O., 1954, Geology of the Golden-Green Mountain area, Jefferson County, Colorado: Colorado School of Mines Quarterly, v. 49, p. 3-96.

Robertson, D., and K. Forrest, 2010, Denver's Street railways, volume 3: The interurbans: Colorado Railroad Museum, Golden, CO, 373 p.

Rockwell, A.P., 1872, Discovery of the tusk of an elephant in Colorado: American Journal of Science, Series 3, v. 3, p. 302.

Rousseau, J.P., 1980, Groundwater hydrology of South Table Mountain, Jefferson, County, Colorado: CSM Thesis (unpublished) ER-1982, 238 p.

Schneider, R., 1984, Coors Rosebud Pottery: Busche-Waugh-Henry Publications, Seattle, 150 p.

Schwochow, S.D., 1978, Proposed North Table Mountain Quarry and Aggregate plants: 23 October 1978. Colorado Geological Survey, unpublished memo, *cited in* P.D. Kilburn and W.L. White, North Table Mountain: Its History and Natural Features https://www.schweich.com/ntm921.html Last accessed 1 June 2021.

Schwochow, S.D., 1980, The effects of mineral conservation legislation on Colorado's aggregate industry, *in* Schwochow, S.D., ed., Proceedings of the Fifteenth Forum on Geology of Industrial Minerals: Colorado Geological Survey Resource Series 8, p. 29-41.

Schwochow, S.D., R.D. Shroba, and P.C. Wicklein, 1974, Sand, gravel, and quarry aggregate resources Colorado Front Range counties: Colorado Geological Survey Special Publication 5-A, 43 p., 3 plates.

Scott, G.R., 1972, Geologic map of the Morrison Quadrangle, Jefferson County, Colorado: USGS Map I-790-A. Map scale 1:24000.

Silber, W.L., 2019, The story of silver: How the white metal shaped America and the modern world: Princeton University Press, New Jersey, 328 p.

Simmons, B., 2004, A quick history of Idaho Springs: Western Reflections Publishing Company, Montrose, Colorado, 85 p.

Simmons, B., and J. Ghist, 2014, Arthur Lakes' dinosaur quarries: A pictorial guide: Friends of Dinosaur Ridge, Morrison, Colorado, 25 p.

Simmons, B., K. Honda, and R. Gardner, 2013, Golden's foundation, the famous "Golden Blue Granite" Jeffco's first granite quarry: Historically Jeffco, Issue 34, p. 29-31.

Slader, W. A., 1952, The White Ash Mine 1874-1889, *in* The Colorado Metallurgical Society, Colorado Metallurgical Society best paper award, 1952, Volume 1: Denver, Colorado, unpaginated. Located in the Mining History Archives Reading Room at Arthur Lakes Library.

Sprain, C.J., P.R. Renne, W.A. Clemens, and G.P. Wilson, 2018, Calibration of chron C29r: New high-precision geochronologic and paleomagnetic constrains from the Hell Creek region, Montana: Geological Society of America Bulletin, v. 130, p. 1615-1644.

State of Colorado, 1882, First biennial report of the Colorado State Industrial School for 1881 and 1882: Colorado Transcript Publisher, Golden, Colorado, 40 p.

State of Colorado, 1906, Thirteenth biennial report of the Colorado State Industrial School for 1905 and 1906: Industrial School Press, Golden, Colorado, 79 p.

State of Colorado, Division of Water Resources, 2019, Water-well database for Colorado: GIS information available at https://dwr.colorado.gov/services/data-information/gis Last accessed 14 June 2021.

Sterne, E.J. (Ned), 2006, Stacked, evolved triangle zones along the southeastern flank of the Colorado Front Range: Mountain Geologist, v. 43, p. 65-92.

Sterne, E.J. (Ned), 2017, Unpublished handouts for field trip, September 2017.

Thwaites, 1905, Early western travels 1748-1846: part II of James's account of S.H. Long's expedition 1819-1820: Arthur H. Clark Company, Cleveland Ohio, p. 279-281.

Tudor Engineering, 1987, Clear Creek Project Phase 1-Feasibility study, Final Report-Appendixes: Colorado Water Resources and Power Development Authority, Denver, 194 p. https://mountainscholar.org/bitstream/handle/10217/21502/WIVG02903.pdf?sequence=1&isAllowed=y Last accessed 14 June 2021

U.S. Department of Interior, 1992, National Park Service National register of historic places: Camp George West. https://npgallery.nps.gov/GetAsset/3fc4d213-cf83-4653-906a-ac60f0cbd7dc Last accessed 14 June 2021.

Upper Clear Creek Watershed Association, 2014, Upper Clear Creek watershed plan update: Clear Creek Consultants and Matrix Design Group, 77 p. https://static1.squarespace.com/static/53f664ede4b032c1fade347d/t/5cc7252adc4c47000182fa6f/1556555089774/CCWatershedPlan_04-07-14_Final.pdf Last accessed 14 June 2014.

Upper Creek Watershed Association, 1997, State of the Watershed Report, Clear Creek 1997: Clear Creek Association, 73 p. with maps. https://static1.squarespace.com/static/53f664ede4b032c1fade347d/t/5cc724b2eef1a1ea6575a6dd/1556554985668/CLEAR-CREEK-State-of-the-Watershed-Report-1997-web.pdf Last accessed 14 June 2021.

USGS Core Research Center, 2019, Analytical data for the Snyder Burbach 20-3H: file D967data17: https://my.usgs.gov/crcwc/core/report/11964 Last accessed June 24, 2019.

USGS Earthquake Catalog, 2021, : https://earthquake.usgs.gov/earthquakes/search/ Last accessed 14 June 2021.

USGS Geographic Names Information System database: https://usgs.maps.arcgis.com/apps/webappviewer/index.html?id=5db1775775694a148df20465605537d4# Last accessed 14 October 2021.

USGS Quaternary Fault Database, 2021, https://earthquake.usgs.gov/cfusion/qfault/show_report_AB_archive.cfm?fault_id=2324&section_id= Last accessed 13 October 2021.

Van Horn, R, 1972, Surficial and bedrock geologic map of the Golden, Quadrangle, Jefferson County, Colorado: USGS Map I-761-A. Map scale 1:24000.

Van Horn, R., 1976, Geology of the Golden Quadrangle: USGS Professional Paper 872, 116 p.

Van Sant, J.N., 1959, Refractory clay deposits of Colorado: U.S. Bureau of Mines Report of Investigation 5553: U.S. Government Printing Office, Washington, D.C., 156 p.

Waage, K. M., 1961, Stratigraphy and refractory clayrocks of the Dakota Group along the northern Front Range, Colorado: USGS Bulletin 1102, 154 p.

Wagenbach, L., 1973, A century of military training at the Colorado School of Mines: Unpublished essay and updates from the Military Science (ROTC) Department at Colorado School of Mines, 4 p.

Warden, B., 2012, Funiculars of Golden, Colorado: B. Warden, Golden, Colorado, 40 p.

Weimer, R.J., 1996, Guide to the petroleum geology and Laramide orogeny, Denver Basin and Front Range, Colorado: Colorado Geological Survey, Bulletin 51, 127 p.

Weimer, R.J., and L.W. Leroy, 1987, Paleozoic-Mesozoic section: Red Rocks Park, I-70 road cut, and Rooney Road, Morrison area, Jefferson County, Colorado: Geological Society of America Centennial Field Guide 1987, p. 315-319.

Weimer, R.J., and R.A. Erickson, 1976, Lyons Formation (Permian), Golden-Morrison area, Colorado: Professional Contributions of Mines, no. 8, p 123-138.

Weimer, R.J., and R.R. Ray, 1997, Laramide mountain-flank deformation and the Golden Fault zone, Jefferson County, Colorado, *in* Bolyard, D.W., and S.A. Sonnenberg, eds., Colorado Front Range Guidebook: Rocky Mountain Association of Geologists, Denver, CO, p. 49-64.

Western Mountaineer, as cited in footnotes. Available at https://www.coloradohistoricnewspapers.org/

Willits, W.C., 1878, Map of Golden, compiled from official records: Denver, Colorado, Map scale 1:3000. https://digital.denverlibrary.org/digital/collection/p16079coll39/id/17/rec/4 Last accessed 14 October 2021.

Willits, W.C., 1899, Willits farm map, including Weld, Boulder, Arapahoe, and Jefferson counties. Map scale approximately 1:39,600: https://digital.denverlibrary.org/digital/collection/p16079coll39/id/394/ Last accessed June 8, 2021.

Yale Peabody Museum (YPM): Correspondence, #B03F0106: Box 1, Folder 106, Berthoud 1861 to 1883, Letter 4, June 21, 1874. https://peabody.yale.edu/collections/vertebrate-paleontology/correspondence-o-c-marsh Last accessed 14 June 2021.

Yehle, L.A., 2001 and various dates, Unpublished correspondence and maps concerning the routes of the Golden City and South Platte Railroad in Golden: Archives of the Colorado Railroad Museum, Golden, Colorado.

Zeigler, V., 1917, Foothills structure in Northern Colorado: The Journal of Geology, v. 25, no. 8, p 715-740.

About the Authors

Donna Anderson

Retired from the oil and gas industry after 40+ years, Donna continues to indulge her geologic passion at the Colorado School of Mines, where she has been a faculty member in the Department of Geology and Geological Engineering since 2000. A native of southern California, Donna earned a B.A. in Earth Science and Geography (double majors) at California State University at Fullerton (1974), an M.S. in Geology at UCLA (1980), and a Ph.D. in Geology at Colorado School of Mines (1997). Donna has never met a rock she didn't like, with the possible exception of those in the Spearfish Formation of Canada and North Dakota.

Donna's career began as an engineering geologist in southern California (1974-1978), where she learned the ropes of geologic hazards analysis for urban development and nuclear power-plant siting. Moving to Colorado, she became a petroleum geologist at Mobil Oil Corp. in Denver (1980-1992), an oil industry consultant (1997-2006), and a geologic advisor at EOG Resources (2006-2015). Times of industry downturns were used as opportunities for re-education and indulging in recreational passions of backpacking and trekking.

At CSM, Donna taught sedimentary geology for petroleum engineers for five years, field camp for petroleum engineers for nine years, and co-taught field camp for geologists for four years. One of her passions is teaching geology to non-geologists, helping them become aware of the world around them without the jargon. She has also taught graduate courses in sedimentary geology, always stressing how to make consistent observations. Her academic research ranges from applied to esoteric.

Giving back to one's profession and community is also a passion for Donna. She is a past-president of several geologic societies, also serving on boards and committees in various capacities. She has received Distinguished Service awards from the Rocky Mountain Association of Geologists (RMAG), the American Association of Petroleum Geologists (AAPG), and the Rocky Mountain Section of AAPG. She is an honorary member of RMAG and AAPG. Donna is a 2014 graduate of Leadership Golden and helped guide that program for two years after her graduation, which is where she first met co-author Paul Haseman. Writing this book is a give-back to the community, inspired by her work with Leadership Golden.

Donna is married to a "dirt" geologist (specializes in Quaternary geology), Larry, now retired from the U.S. Bureau of Reclamation, who spent his career in earthquake hazards analysis for dams and other critical structures in the Western U.S. Larry volunteers at the Colorado Railroad Museum, where he indulges his passion for railroads and history, including a model railroad in the basement of their home. She and Larry have lived in Colorado since 1980 and in Golden since 1984, always with a designated dog, currently a Golden Retriever appropriately named Dakota, after the rock formation.

Paul Haseman

As a retired attorney, Paul keeps his hand in with several Golden civic groups such as Leadership Golden, Golden United, Rotary, Lions and a member of the Golden City Council. With obvious time on his hands, a couple years ago Paul contacted a geologist acquaintance to discern her interest to join him in writing about the Geology of Golden. Little did he know that Donna Anderson keeps her Ph.D. office at the School of Mines and has published many geology papers. She jumped at the chance, and you are reading the result.

Paul's interest in Geology springs for a couple sources, most notably from his father, Leonard Haseman, who was asked to advise NASA on the thickness of dust before the first Moon landing, based on his graduate studies with the Corps of Engineers, at Cornell. Colonel Haseman also was the first head of the Corps' Geodesy Intelligence Research and Development Agency, which first mapped the "dark" side of the Moon. An inveterate "rockhound," he gave all seven of us children periodic geology lessons on rock formation during our many cross-country camping trips. Thus, love of "rocks" became deeply imbedded in our genes.

More recently Paul's handy copy of the "Roadside Geology of Colorado" and several trips to the Mines Museum sparked his interest in the renown geology of Golden. So, let's write a book but not any book, rather a book the residents of Golden would enjoy reading. Paul's military training (LTC, US Army Ret) taught him the acronym, KISS – keep it simple, stupid. Donna and Paul have worked hard to suppress a pedantic tone and keep it simple and interesting, while still scholarly and accurate. We hope you agree.

The Colorado School of Mines also agreed with our KISS concept and is publishing this book on-line with free access to every-day residents and perhaps a few students. Print copies have been provided to all Golden schools. Young and old, we hope you find Golden's unique geology and mining history very special indeed.

Paul holds degrees from USMA/West Point (BS), University of Virginia (JD) and George Washington University (LLM). Following his military service in the Corps of Engineers and as a Judge Advocate General (JAG), Paul worked in aerospace with various positions to include General Counsel, Raytheon Australia. He authored a book, <u>Mule Memories</u>, covering his years at West Point and wrote a weekly newspaper column for 13 years, while living in California, where he also served on the Capistrano Unified District School Board for 17 years.

He and Vivian have been married 52 years with three children and 12 grandchildren and their 15 year-old miniature poodle, Coco. Paul and Vivian have lived in Golden for eight years.